MOST®
Work Measurement Systems

INDUSTRIAL ENGINEERING

A Series of Reference Books and Textbooks

Editor

WILBUR MEIER, JR.
Head, School of Industrial Engineering
Purdue University
West Lafayette, Indiana

Volume 1: Optimization Algorithms for Networks and Graphs, *Edward Minieka*

Volume 2: Operations Research Support Methodology, *Albert G. Holzman*

Volume 3: MOST Work Measurement Systems, *Kjell B. Zandin*

Additional Volumes in Preparation

MOST®
Work Measurement Systems

KJELL B. ZANDIN

H. B. Maynard and Company, Inc.
Pittsburgh, Pennsylvania

MARCEL DEKKER, INC., New York and Basel

Library of Congress Cataloging in Publication Data

Zandin, Kjell B. [date]
MOST® work measurement systems.

(Industrial engineering; v. 3)
Includes index.
1. Work measurement. I. Title. II. Series.
T60.Z36 658.5'42 80-10628
ISBN 0-8247-6899-X

MARCEL DEKKER, INC.
270 Madison Avenue, New York, New York 10016

Current printing (last digit):
10 9 8 7

PRINTED IN THE UNITED STATES OF AMERICA

When you can measure what you are speaking of and express it in numbers you know that on which you are discoursing. But if you cannot measure it and express it in numbers, your knowledge is of a very meagre and unsatisfactory kind.

Lord Kelvin

Measure of work brings knowledge. Through this knowledge, factual decisions and improvements can be made and control exercised.

This text is unquestionably intended to give the reader a complete description of MOST Work Measurement Systems. It is not, however, the sole training device through which MOST Systems is learned. Any attempt to utilize the material in this text without proper classroom training and certification will be done at the discretion of the reader.

Through Chapters 2 (The General Move Sequence), 3 (The Controlled Move Sequence), 4 (The Tool Use Sequence), and 7 (Application of MOST Work Measurement Technique), this text is intended to give the reader a complete understanding of the basic MOST Work Measurement technique and its application to industrial situations. To a certain degree the Equipment Handling Sequences (Chapter 5) are described in some detail; however, certain parameters must be validated prior to their application. The scope of MOST Clerical Systems (Chap. 6), MOST Application Systems (Chap. 8), and MOST Computer Systems (Chap. 9), would dictate that adequate coverage of these subjects could well be addressed in separate texts. With that in mind, Chapters 6, 8, and 9 are included in this volume for information purposes only.

Contents

Foreword

Work measurement provides a systematic methodology for establishing the amount of time it should take the human operator to perform a given task. Measurement of work is fundamental to accomplishing production, planning, and management. In fact, work measurement is to the production system designer what surveying is to the civil engineer.

Because it is costly, unnecessarily time consuming, and often impossible to develop time standards for particular production tasks, engineers and their supporting technical staff resort to the use of predetermined time systems to establish the time required for completing tasks primarily in manufacturing and processing industries. A variety of predetermined time systems have been developed over the past three to four decades, each with specific purposes and objectives.

H. B. Maynard and Company, Inc. has been a leader in the development and education in the use of these systems. One of the popular systems, Methods Time Measurement (MTM), was developed by Dr. Harold B. Maynard, the founder of the company. This system has been modified and in some cases simplified over the years and represents one of the most widely used systems available.

H. B. Maynard and Company is now making available a new predetermined time system, the Maynard Operation Sequence Technique (MOST), which was developed by Kjell B. Zandin and tested extensively by the company. MOST, the subject of this book, is a novel system which achieves a simplification in system application for a variety of process tasks while not sacrificing the accuracy of the results. The MOST System was developed recognizing that no two individuals perform the same tasks in the identical way. Thus, work performed by different individuals varies from one individual to another. As time standards are *average* times for completing a task, this natural variation can be exploited to develop a technique requiring fewer elements to construct a time standard. These elements are actually sequences of elemental tasks. The MOST technique is built around only three basic sequences, for general spatial move, controlled move, and tool use.

This book and the MOST technique provide needed additions to the industrial engineering literature. An ever-increasing need for productivity improvements in manufacturing and service industries, coupled with rapidly changing production tasks, creates an environment in which predetermined time standards that can be more conveniently used will be increasingly desirable.

Wilbur Meier, Jr.
Head, School of Industrial Engineering
Purdue University

Introduction

One of the great rewards of any profession is the pleasure of being involved in an innovative development that has great potential benefits for the profession and for the industrial world. Countless instances of such developments have occurred in all professions over the years. In the history of the industrial engineering profession, H. B. Maynard and Company has been particularly fortunate in this respect. Our founder, H. B. (Mike) Maynard and his close friend and collaborator, G. J. (Gus) Stegemerten were two of the great innovators in the history of the industrial engineering profession. Together, they developed many widely used industrial engineering techniques such as methods time measurement, skill and effort leveling factors, operations analysis, and universal maintenance standards.

When methods time measurement was first introduced publicly, its acceptance was eagerly anticipated throughout H. B. Maynard and Company. The professional staff was well aware of the potential importance of this development and the tremendous impact it could have on improving productivity throughout industry, business, and government. As MTM gained more and more acceptance there was a deep feeling of accomplishment not only by Mike Maynard and Gus Stegemerten, but by everyone in the firm who participated in its development and introduction.

For the last few years the professional group at H. B. Maynard and Company has been experiencing anew that same sort of feeling with the introduction of MOST Systems. Starting in the United States in 1975, MOST Systems has gained wide recognition as a major contribution to the body of industrial engineering technology. In the first five years since MOST Systems has been available in the United States, literally hundreds of organizations and thousands of individuals have been trained to use MOST. A number of leading industrial engineers throughout the world have acclaimed MOST Systems as the wave of the future in industrial engineering.

The development of MOST Systems from a base of methods time measurement is the result of a continuing evolution in industrial engineering technology. In 1940, the Methods Engineering Council (the former name of H. B. Maynard and Company) conducted a supervisory training program at a large plant of one of the major United States corporations. The training revolved around the role of the supervisor in improving shop productivity, and included practical application of work simplification techniques to shop operations. Many substantial cost reduction ideas were generated during the training program, and the program was considered to be quite successful, from a monetary standpoint.

In reviewing the results of the program, however, Maynard and Stegemerten questioned whether the program was completely successful because of the upsetting effect that the many changes caused in the plant. The training resulted in a very strong "methods correction" drive on the part of the foreman that was very fruitful, but also very controversial because of the resistance to change that had to be overcome in the plant. They reasoned that if the industrial engineers had access to a better "methods-engineering" tool to set up operations correctly in the first place, then subsequent methods correction could be minimized, thus also minimizing the negative effects of changes within an organization.

This basic philosophy sent Maynard and Stegemerten on a research course to find a better way to engineer methods. With the assistance of J. L. Schwab and others, they did, indeed, find a better way. They created methods time measurement (MTM), which rapidly became one of the best known and most widely used work study systems. Its use has saved many billions of dollars in improved productivity and reduced costs.

One of the philosophies that Maynard advanced was that "with sufficient study any method can be improved." Following this philosophy, industrial engineers have been seeking improvements over methods time measurement and other predetermined motion time systems for several decades. These efforts have resulted in a considerable array of predetermined data systems, like MTM-2, MTM-3, MSD, USD, GPD, MTMV, and others. The objective, of course, was to retain the good features of MTM for analyzing methods and measuring work, but to eliminate the handicap of lengthy application times inherent in the MTM system. Each of these evolving systems has been successful, to a degree, in meeting the objective of reducing the industrial engineering application time, and each system has had its advocates as the best such second-level MTM system to be used.

In the late 1960s, Kjell Zandin, then working for the Swedish Division of H. B. Maynard and Company, made an important discovery. While reviewing the rather extensive MTM data in the Maynard library in Gothenburg he detected striking similarities in the sequence of MTM defined motions whenever any object was handled. Invariably, the same set of basic motions would be used in the same general sequence. This discovery led Zandin and the Maynard management to question whether this phenomenon could be used to develop a new way to analyze methods and to measure operations. If successful, this approach could drastically reduce the time required to study operations.

For the next several years Zandin spent the bulk of his time conducting intensive research in the development of this new concept of work study. He isolated and developed models for three motion sequences that would analyze and measure practically all manual work. Later, he identified three other sequences which would measure practically all heavy material handling that required mechanical assistance. Again, relying on Maynard MTM library data and statistical methods, he developed a set of several index numbers to be used with the sequence models. After the application procedures were spelled out, Zandin and other Maynard personnel made elaborate tests of the system and data in a variety of industries in Sweden and Western Europe.

To the great satisfaction of all concerned, the new system worked remarkably well. Without question, it represented a very significant advance in the state of the art in industrial engineering. It was fast (forty to fifty times faster than MTM-1), accurate, methods conscious, and easy to learn and apply. The new system was called Maynard Operation Sequence Technique (MOST) and was ready for wide distribution in 1975.

There is a solid future for MOST Systems throughout the world. Still in its infancy, MOST has been applied successfully in practically all industries, ranging in diversity from shipbuilding to electronics; truck assembly to textiles; freight car assembly to drugs; and furniture to food products. Applications have been made in offices, assembly shops, fabrication and welding shops, production lines, job shops, material handling, maintenance, warehouses, and finishing operations. In fact, the experience to date has shown that MOST Systems are truly universal in their application throughout industry, business, and government.

Continuing to evolve MOST Systems have been developed into MOST Computer Systems, as explained in Chapter 9 of the book. MOST Computer Systems have incorporated all of the benefits of the MOST Manual System, plus the great advantage of mass data updating, speed of application, data development capabilities, and rapid access to file information. With the constant improvement of computer technology, we anticipate that MOST Systems will continue to evolve to meet the unceasing demand for finding better ways to do things.

William M. Aiken
H. B. Maynard and Company, Inc.

Acknowledgments

It is not a one-man job to create a new work measurement system that will help thousands of industrial engineers, the users, do their work better and with more satisfaction. After several years of struggle, which resulted in a more conventional type of system for machining operations, I was fortunate enough to hit upon a new concept in the field of measuring manual work.

Based on fundamental statistical principles and basic work measurement data compiled over many years, the idea of MOST evolved as the most natural thing in the world. At all stages of the development of MOST the practical aspects were consistently emphasized. The goal was to make MOST a useful industrial engineering tool, easy to learn and simple to apply.

MOST cropped up late one afternoon in August 1967. The concoction hit the desk in a jammed layout room as a ripe apple hits the ground. Then, it was inconceivable that MOST would "conquer" the world a few years later. The "baby" did not even have a name. It was not until 1975 that "MOST" became MOST.

Many people have been very helpful and supportive in forming MOST. Their contributions and encouragement have been highly appreciated and invaluable in making MOST a contemporary "star" in the work measurement world.

My lovely wife, Sonja, has always been my greatest supporter. I am most grateful to her.

MOST would never have gotten off the ground without the vision, the back-up, and the total faith of Lennart Gustavsson, Managing Director of Maynard MEC in Gothenburg, Sweden. I owe my warmest thanks to my friend, Lennart.

Without the foresight and business mind of William M. Aiken, President of H. B. Maynard and Company in Pittsburgh, MOST would never have been successfully launched in the United States. This book might not have been written without Bill's initiative to expose MOST to U.S. industry and government. For giving me the challenging opportunity to move across the Atlantic together with my family six years ago, my sincerest thanks to Bill. It has been an exciting adventure and an extremely educational and rewarding experience.

It took many years to polish, refine, and finalize the back-up data for MOST. Numerous MTM analyses had to be made, tested, and validated. My colleague and good friend, Thomas Vago, from Lund, Sweden, has persistently assisted me with the "nitty gritties" to make the foundation for MOST totally solid and unassailable. Thomas's competence and interest have been of great value.

By utilizing just a small portion of his extensive knowledge, Dr. William D. Brinckloe, Professor of Public Administration and International Affairs at the University of Pittsburgh, intelligently and explicitly clarified and confirmed the statistical theories behind MOST. Dr. Brinckloe was the first to emphasize the unique consistency of MOST. My thanks to him for his professional and sincere help.

One of the most demanding and time-consuming tasks in completing this book has been to ensure that the text is logical and complete, the examples illustrative, well defined, and representative of a variety of common industrial activities. William M. Yates, Jr., Technical Coordinator of MOST Systems with H. B. Maynard and Company in Pittsburgh, has done his utmost to streamline and clarify the text. Thanks to Bill's positive criticism and dedicated work, the quality of the book has improved immensely. Bill deserves my sincerest thanks.

Mary T. Coughlin has been very instrumental in substantiating the happy marriage between MOST and a minicomputer, and Ronald A. Soncini has devotedly and enthusiastically brought MOST into the office for measuring clerical activities. Many others, Stig Magnusson, Lennart Simrén, Fred Berglund, Berndt Nyberg, and Bob Hooks, just to mention a few, have contributed with their knowledge and efforts to make MOST a practical tool for the industrial engineer. Their work to advance and expand MOST is sincerely appreciated.

All sketches, diagrams, and tables have been skillfully drawn by the artistic hand of Bette J. McDonald. The text was untiringly typed and retyped by my secretary, Jacqueline D. Flaherty, and proofread by Bob Dietrich.

Finally, I owe a great deal to all the industrial engineers who are daily practicing MOST and who have expressed their satisfaction and encouragement. Based on their appreciation, I am now totally confident that MOST will become a "way of life" for the modern industrial engineer. It already is for some.

Kjell B. Zandin

1
The Concept of MOST—An Introduction

Work Measurement

The desire to know how long it should take to perform work must surely have been present in those individuals responsible for erecting ancient monuments or shaping tools. Why did the ancients and why do we need to be able to predict with accuracy the length of a working cycle? How was such a prediction made? How is it made now?

There are many reasons for wanting to know the amount of time a particular task should take to accomplish. It may simply be for reasons of curiosity. But, realistically, it would be for any of three reasons: to accomplish planning, to determine performance, and to establish costs. Suppose an organization wished to manufacture a new product. Using an economical, predetermined, motion time system, the planning and budgeting process could be accomplished. Knowing the time to manufacture and assemble various parts and/or components, a manager could:

- Determine the total labor cost of the product.
- Determine the number of production workers needed.
- Determine the number of machines needed.
- Determine the amount of and delivery times for material.
- Determine the overall production schedule.
- Determine the feasibility of entering into production of the product.
- Set production goals.
- Follow up on production: Have goals been achieved?
- Check individual or departmental efficiency.
- Know the actual costs of production.
- Pay by results.

As a consequence, a manager can achieve an even and sufficiently high utilization of personnel, material, and equipment to result in an overall efficiency that will allow an organization to survive and grow.

It must be that the original form of work measurement was guessing. It is interesting to note that the primitive guessing technique employed thousands of years ago is still in use today in many modern manufacturing organizations. Today's version is a much advanced form of the original technique, however, and is known as an educated guess. The educated guess is unscientifically supported by intuition, individual personal experience, the importance of the estimation to be made, and the inherent ability or inability of the applicator to make a confident-sounding response. Obviously, this technique is not scientific (well documented or statistically supported) and not accurate (with any degree of confidence or consistency), but it is fast.

Once products began to be manufactured or work tasks completed, another source of information was available from which future times could be estimated. The historical data concept of work measurement evolved. From records of what had been accomplished came the information for predicting times for future situations. Using historical data does one thing very well. It accurately tells you what has already happened. To use it to predict what will happen assumes two major points,

1. The conditions and actions under which the process was performed originally are what you wish to repeat (are the best way of performing a task).
2. The actions to be performed will be performed exactly as those on which the historical data are based.

If these conditions are met, historical data should work well.

A true innovator, Frederick Taylor, looked at work as something that could be controlled or engineered. It did not have to be a haphazard repetition of what had gone on before; in fact, workers could be instructed as to the best way to perform certain tasks. The result was that tasks were broken down into elements or short tasks that could be arranged and managed to produce more efficient and productive and less fatiguing work. Each element was studied to determine which was productive and which was useless. Keeping only productive elements, a stopwatch was used to determine the time for each. The time recorded was the actual time taken by a particular individual to perform a certain task under specific conditions. To make such times transferable to other workers and other situations, time for the average worker working under average conditions had to be determined. This was and is now accomplished by performance rating. The performance rating is a determination by the analyst of the pace of the individual observed as compared to the ideal, imaginary, average worker working at a level of 100% effort and skill. If the worker observed is not putting forth the effort imagined to be 100%, a rating of less than 100% would be applied to the time recorded by the stopwatch and the time would be leveled to 100% performance. Likewise, if the worker observed was working with more skill and effort than the imagined average worker, a rating of over 100% would be applied to the time on the stopwatch and the time leveled to 100% performance. The scientific process of engineering a task using the time study methods just described has two weak points:

1. The individual analyst must subjectively rate or compare the operator to an estimated 100% performance standard.
2. No matter how sophisticated, expensive, or precise the timepiece, a watch simply does not forecast, predict, or accurately determine times for future situations; it only determines what has already occurred.

It was discovered by Frank and Lillian Gilbreth that all manual operations were combinations of basic elements. The Gilbreths isolated and identified these elements primarily so that methods could be more accurately explained and improved. They reasoned that to reduce the motion content of a task was to reduce the effort and the time to perform the task. The result is higher production.

Understandably, followers of Taylor practiced time study, while followers of the Gilbreths practiced motion study. As frequently occurs, a third party entered and joined together the best of both techniques. A marriage of the time study technique and the motion study philosophy was arranged. From this union of time and motion studies was born the predetermined motion time systems (PMTS). These systems utilized the time study and micromotion techniques of the earlier techniques to determine and assign times to specified basic motions. The motions and associated times were cataloged. Work measurement then became a matter of establishing the best basic motion pattern to perform a certain task, and from the catalog or data card, assigning the appropriate predetermined time for each basic motion in that pattern. Since the times for all motions are predetermined, one could now accurately predict future task times. The watch was needed only for timing machines. But what about performance rating? The authors of the best predetermined motion time systems built into their systems the leveled times for 100% performance. Therefore, with the catalogs of predetermined times already leveled to 100%, there no longer remained a need to rate an operator. The analyst began to focus on the actual work being accomplished, not on the operator.

Of all the predetermined motion time systems, the most well known is methods time measurement (MTM), as developed by Harold B. Maynard, G. J. Stegemerten, and J. L. Schwab and published in 1948. Being a very detailed system and being in the public domain, MTM has been recognized as the most accurate and widely accepted predetermined motion time system in use today.

The MTM system has a detailed data card of basic motions (reach, move, grasp, position, release, body, leg and foot motions, and so on), each concerned with particular variables. Basic motions are identified, and with the variables considered the appropriate times are chosen from the data card. Because of its detail, MTM can be a very exact system and also very slow to apply. Also, basic motion distances must be accurately measured and correctly classified. Because of the detail, applicator error can be a problem. The times which result from performing an MTM analysis do reflect a 100% performance level, and times can be established for operations prior to production.

Synthesized versions of MTM were developed to reduce applicator error and the time of analysis. Two such versions are MTM-2 and MTM-3. These systems grouped or averaged together certain basic motions and/or variables to reduce the applicator effort required to apply the technique. A corresponding lack of system accuracy also resulted (see Appendix A, Theory).

The analysis of work, as practiced by industrial engineers using a predetermined motion time system today, is performed by systematically breaking work down into very small and distinct units called basic motions. For highly repetitive, short-cycle operations, this attention to detail is usually necessary and, indeed, has been found to be quite effective in generating valuable methods improvements. For less repetitive operations or job shop production, however, this detailed approach is very tedious and requires a great deal of time and effort on the part of highly trained engineers and technicians. The benefits are often questionable when considering the amount of analysis effort required. It can be a very costly process.

The Concept of MOST Work Measurement Technique

Because industrial engineers are trained that with sufficient study any method can be improved, many efforts have been made to simplify the analyst's task. This has, for instance, led to a variety of higher-level MTM data systems now in use. This attitude also led us to examine the whole concept of work to find a better way for analysts to accomplish their mission. This was the formation of the concept later to be known as MOST, Maynard Operation Sequence Technique.

Work to most of us means exerting energy, but we should add, to accomplish some task or to perform some useful activity. In the study of physics, we learn that work is defined as the product of force times distance ($W = f \times d$) or, more simply, work is the displacement of a mass or object. This definition applies quite well to the largest portion of the work accomplished every day (e.g., pushing a pencil, lifting a heavy box, or moving the controls on a machine). Thought processes or thinking time is an exception to this concept, as no objects are being displaced: For the overwhelming majority of work, however, there is a common denominator from which work can be studied, the displacement of objects. All basic units of work are organized (or should be) for the purpose of accomplishing some useful result by simply moving objects. That is *what* work is. MOST is a system to measure work; therefore, *MOST concentrates on the movement of objects.*

Work, then, is the movement of objects, maybe we should add, following a tactical production outline. Efficient, smooth, productive work is performed when the basic motion patterns are tactically arranged and smoothly choreographed (methods engineering). It was noticed that the movement of objects follows certain consistently repeating patterns, such as reach, grasp, move, and position the object. These patterns were identified and arranged as a sequence of events (or subactivities) followed in moving an object. A model of this sequence is made and acts as a standard guide in analyzing the movement of an object. It was also noted that the actual motion content of the subactivities in that sequence vary independently of one another.

This concept provides the basis for the MOST Sequence Models. The primary work units are no longer basic motions, but fundamental activities (collections of basic motions) dealing with moving objects. These activities are described in terms of subactivities fixed in sequence. In other words, to move an object, a standard sequence of events occurs. Consequently, the basic pattern of an object's movement is described by a universal sequence model instead of random, detailed basic motions.

Objects can be moved in only one of two ways: either they are picked up and moved freely through space, or they are moved and maintain contact with another surface. For example, a box can be picked up and carried from one end of a workbench to the other or it can be pushed across the top of the workbench. For each type of move, a different sequence of events occurs; therefore, a separate MOST activity sequence model applies. The use of tools is analyzed through a separate activity sequence model which allows the analyst the opportunity to follow the movement of a hand tool through a standard sequence of events, which, in fact, is a combination of the two basic sequence models.

Consequently, only three activity sequences are needed for describing manual work. The MOST technique therefore is comprised of the following basic sequence models:

- The General Move Sequence (for the spatial movement of an object freely through the air)
- The Controlled Move Sequence (for the movement of an object when it remains in contact with a surface or is attached to another object during the movement)
- The Tool Use Sequence (for the use of common hand tools)

The Basic MOST Sequence Models

General Move is defined as moving objects manually from one location to another freely through the air. To account for the various ways in which a General Move can occur, the activity sequence is made up of four subactivities:

A Action distance (mainly horizontal)
B Body motion (mainly vertical)
G Gain control
P Place

These subactivities are arranged in a Sequence Model (Table 1.1).

These subactivities, or sequence model parameters as they are called, are then assigned time-related index numbers based on the motion content of the subactivity. This approach provides complete analysis flexibility within the overall control of the sequence model. For each object moved, any combination of motions could occur, and, using MOST, any combination could be analyzed. For the General Move Sequence, these index values are easily memorized from a brief data card (Table 2.1). A fully indexed General Move Sequence, for example, might appear as follows:

$A_6 \quad B_6 \quad G_1 \quad A_1 \quad B_0 \quad P_3 \quad A_0$

Table 1.1 Sequence Models Comprising the Basic MOST Technique

MANUAL HANDLING		
ACTIVITY	SEQUENCE MODEL	SUBACTIVITIES
GENERAL MOVE	ABGABPA	A - ACTION DISTANCE B - BODY MOTION G - GAIN CONTROL P - PLACE
CONTROLLED MOVE	ABGMXIA	M - MOVE CONTROLLED X - PROCESS TIME I - ALIGN
TOOL USE/EQUIPMENT USE*	ABGABP ABPA	F - FASTEN L - LOOSEN C - CUT S - SURFACE TREAT R - RECORD T - THINK M - MEASURE

*The Equipment Use Sequence Model and its appropriate subactivities are an integral part of the MOST Clerical Systems and are therefore discussed in a later chapter.

where A_6 = Walk three to four steps to object location
B_6 = Bend and arise
G_1 = Gain control of one light object
A_1 = Move object a distance within reach
B_0 = No body motion
P_3 = Place and adjust object
A_0 = No return

This example could, for instance, represent the following activity: walk three steps to pick up a bolt from floor level, arise, and place the bolt in a hole.

General Move is by far the most frequently used of the three sequence models. Roughly 50% of all manual work occurs as a General Move, with the percentage running higher for assembly and material-handling work, and lower for machine shop operations.

The second type of move is described by the Controlled Move Sequence (Table 1.1). This sequence is used to cover such activities as operating a lever or crank, activating a button or switch, or simply sliding an object over a surface. In addition to the A, B, and G parameters from the General Move Sequence, the sequence model for controlled move contains the following subactivities:

M Move controlled
X Process time
I Align

As many as one-third of the activities occurring in machine shop operations may involve Controlled Moves. In assembly work, however, the fraction is usually much smaller. A typical activity covered by the Controlled Move Sequence is the engaging of the feed lever on a milling machine. The sequence model for this activity might be indexed as follows:

$$A_1 \quad B_0 \quad G_1 \quad M_1 \quad X_{10} \quad I_0 \quad A_0$$

where A_1 = Reach to the lever a distance within reach
B_0 = No body motion
G_1 = Get hold of lever
M_1 = Move lever $\leqslant$ 12 in. (30 cm) to engage feed
X_{10} = Process time of approximately 3.5 seconds
I_0 = No alignment
A_0 = No return

The third sequence model comprising the basic MOST technique is the Tool Use or Equipment Use Sequence Model. This sequence model covers the use of hand tools for such activities as fastening or loosening, cutting, cleaning, gauging, and writing. Also, certain activities requiring the use of the brain for mental processes can be classified as Tool Use, e.g., reading and thinking. As indicated above, the Tool Use Sequence Model is a combination of General Move and Controlled Move activities. It was developed as a part of the basic MOST Systems, merely to simplify the analysis of activities related to the use of hand tools. It will later become obvious to the reader that any hand tool activity is made up of General and Controlled Moves.

The use of a wrench, for example, might be described by the following sequence:

$$A_1 \quad B_0 \quad G_1 \quad A_1 \quad B_0 \quad P_3 \quad F_{10} \quad A_1 \quad B_0 \quad P_1 \quad A_0$$

where A_1 = Reach to wrench
B_0 = No body motion
G_1 = Get hold of wrench
A_1 = Move wrench to fastener a distance within reach
B_0 = No body motion
P_3 = Place wrench on fastener
F_{10} = Tighten fastener with wrench
A_1 = Move wrench a distance within reach
B_0 = No body motion
P_1 = Lay wrench aside
A_0 = No return

These three sequence models just described and the subactivites comprising them are presented in Table 1.1.

Time Units

The time units used in MOST are identical to those used in the basic MTM system (methods time measurement), and are based on hours and parts of hours called time measurement units (TMU). One TMU is equivalent to 0.00001 hours. The following conversion table is provided for calculating standard times:

1 TMU = 0.00001 hour
1 TMU = 0.0006 minute
1 TMU = 0.036 second

1 Hour = 100,000 TMU
1 Minute = 1,667 TMU
1 Second = 27.8 TMU

The time value in TMU for each sequence model is calculated by adding the index numbers and multiplying the sum by 10. In our previous General Move Sequence example, the time would be $(6 + 6 + 1 + 1 + 0 + 3 + 0) \times 10 = 170$ TMU, corresponding to 0.1 minute. The time values for the other two examples are computed in the same way. The Controlled Move totals up to $(1 + 0 + 1 + 1 + 10 + 0 + 0) \times 10 = 130$ TMU and the Tool Use, $(1 + 0 + 1 + 1 + 0 + 3 + 10 + 1 + 0 + 1 + 0) \times 10 = 180$ TMU.

All time values established by MOST reflect an average skilled operator's speed at an average performance level or normal pace. This is often referred to as the 100% performance level that in time study is achieved by using "leveling factors" to adjust times to defined levels of skill and effort. Therefore, when using MOST, it is *not necessary* to adjust times unless they must conform with particular high or low task plans used by some companies. This also means that a properly established time standard, by using either MOST, MTM, or stopwatch time study, will give nearly identical results in TMU.

The analysis of an operation will consist of a series of sequence models describing the movement of objects to perform the operation. Total time for the complete MOST analysis is arrived at by adding the computed sequence times. The operation time may be left in TMU or converted to minutes or hours. Again, this time would reflect pure work content (no allowances) at the 100% performance level.

Parameter Indexing

The objective of a work measurement system is to provide for the documentation of a specific work method with the corresponding time. As we have seen, this is accomplished in MOST by applying time-related index numbers to each sequence model parameter, based on the motion content of the subactivity. Parameter indexing, as it is called, is the process of selecting the appropriate parameter variant from a data card or table (Table 2.1, for example) and applying the corresponding index number. With training and practice, parameter variants and index numbers are committed to memory by the MOST analyst. Practically all analysis work can therefore be performed without any direct assistance from data cards or index tables.

Time values for each parameter variant located on the data cards are based on detailed MTM-1 or MTM-2 backup analyses. These MTM analyses are arranged or "slotted" into fixed time ranges represented by an index number corresponding to the median time of each range. The time ranges were calculated using statistical accuracy principles (see Appendix A, Theory).

The Equipment Handling Sequence Models

While the three manual sequences comprise the basic MOST technique, three other sequence models were designed to simplify the work measurement procedure for dealing with heavy objects. These special sequence models (presented in Table 1.2) cover the movement of objects using material-handling equipment.

The Manual Crane Sequence covers the use of a manually traversed jib crane, monorail crane, or bridge crane for moving heavier objects. This sequence

Table 1.2 MOST Sequence Models for Equipment-Handling Objects

EQUIPMENT HANDLING		
ACTIVITY	SEQUENCE MODEL	SUBACTIVITY
MOVE WITH MANUAL CRANE (JIB TYPE)	ATKFVLVPTA	A - ACTION DISTANCE T - TRANSPORT EMPTY K - HOOK UP AND UNHOOK F - FREE OBJECT V - VERTICAL MOVE L - LOADED MOVE P - PLACE
MOVE WITH POWERED CRANE (BRIDGE TYPE)	ATKTPTA	A - ACTION DISTANCE T - TRANSPORT K - HOOK UP AND UNHOOK P - PLACE
MOVE WITH TRUCK	ASTLTLTA	A - ACTION DISTANCE S - START AND PARK T - TRANSPORT L - LOAD OR UNLOAD

is applicable, for example, to machine shop operation where a jib crane is provided at the workplace for loading and unloading the machine. Loading a part into the chuck of an engine lathe might be indexed in the following manner:

$$A_6 \quad T_{16} \quad K_{24} \quad F_3 \quad V_{16} \quad L_{24} \quad V_6 \quad P_{24} \quad T_{10} \quad A_3$$

where A_6 = Walk four steps to 1-ton crane
T_{16} = Obtain crane and transport 9 ft. (2.5 m) to object
K_{24} = Hook up object with single hook (and subsequently unhook it)
F_3 = Free object from pallet
V_{16} = Raise object 45 in. (115 cm)
L_{24} = Transport load 17 ft. (5 m) to machine chuck
V_6 = Lower object 15 in. (40 cm)
P_{24} = Place object in chuck with several adjustments
T_{10} = Aside crane 3 ft. (1 m) away
A_3 = Return two steps to machine

Like the General Move, Controlled Move, and Tool Use Sequences, the time in TMU for completing this jib crane example is computed by adding up all index numbers and multiplying by ten: (6 + 16 + 24 + 3 + 16 + 24 + 6 + 24 + 10 + 3) X 10 = 1320 TMU.

However, for the Powered Crane and Truck Sequences, the conversion to TMU is accomplished by multiplying the sum of the index numbers by 100 instead of 10. This is because the activities described by these sequences cover much larger segments of work. The use of a pendant-controlled bridge crane, for example, might be analyzed in the following way:

$A_1 \quad T_{16} \quad K_6 \quad T_{16} \quad P_3 \quad T_{16} \quad A_1$

where A_1 = Walk 24 ft. (7 m) to crane control panel
T_{16} = Transport crane 25 ft. (18 m) to object
K_6 = Hook up object with single hook and subsequently unhook it
T_{16} = Transport load 25 ft. (18 m) away
P_3 = Place object with single change of direction
T_{16} = Aside crane 25 ft. (18 m) away
A_1 = Return 24 ft. (7 m) to workplace

The time for this activity is (1 + 16 + 6 + 16 + 3 + 16 + 1) X 100 = 5900 TMU. This sequence model is used to cover any type of power traversed crane.

The Truck Sequence covers the transportation of objects using riding or walking equipment such as a forklift, stacker, pallet lift, or hand truck. The use of a forklift to pick up a loaded pallet and move it to a pallet rack, for example, may occur as follows:

$A_1 \quad S_6 \quad T_3 \quad L_6 \quad T_3 \quad L_{10} \quad T_3 \quad A_1$

where A_1 = Walk 24 ft. (7 m) to forklift
S_6 = Start (and subsequently park) forklift
T_3 = Drive forklift 57 ft. (17 m) to pallet
L_6 = Load pallet on forklift from the floor
T_3 = Transport pallet 60 ft. (18 m) to rack
L_{10} = Unload pallet in rack
T_3 = Drive forklift 55 ft. (17 m) to parking area and park
A_1 = Return 24 ft. (7 m) to workplace

The time for this sequence is (1 + 6 + 3 + 6 + 3 + 10 + 3 + 1) X 100 = 3300 TMU.

Application Speed

MOST is generally much faster than other work measurement techniques because of its simpler construction. Predetermined motion time systems are traditionally based on assigning selected time values to minute human motions. For example, to arrive at a time standard for putting a part into a machine, each basic motion involved must be identified, recorded, and assigned time values selected from tables. The values are then added together to arrive at the time for performing the complete operation.

MOST does not require that operations be broken down into such detail. Instead, MOST groups together the basic motions that frequently occur in sequence. Arriving at a standard time with MTM for putting a part into a drill press might require the identification of as many as 15 separate basic motions followed by the assignment of time values to each from the MTM card. Using MOST, the same analysis requires the identification directly from memory of only seven subactivities. Sequence models are preprinted on the analysis form, leaving the analyst with the task of filling in only the variable index numbers.

A comparison between the speed of MOST and other work measurement techniques is shown in Table 1.3, based on the number of measurement units completed per analyst hour. In this study, 1 hour of analyst time yielded 300 TMU of measured work with MTM-1. MTM-2 and MTM-3 yielded 1000 and 3000 TMU, respectively. Using MOST, the same amount of analyst time yielded 12,000 TMU. In other words, MOST was 40 times faster to apply than MTM-1, 12 times faster than MTM-2, and 4 times faster than MTM-3. As a general rule, 1 hour of work can be measured with an average of 10 hours of MOST analyst time. *Note:* The above comparison was performed under laboratory conditions; actual in-plant application may yield total TMU output other than that indicated above.

Accuracy

The accuracy principles that apply to MOST are the same as those used in statistical tolerance control. That is, the accuracy to which a part is manufactured depends on its role in the final assembly. Likewise, with MOST, time values are based on calculations that guarantee the overall accuracy of the final time standard. Based on these principles, MOST provides the means for covering a high volume of manual work with accuracy comparable to existing predetermined

Table 1.3 Comparison of Application Speed

WORK MEASUREMENT TECHNIQUE	TOTAL TMU PRODUCED PER ANALYST HOUR
MTM-1	300
MTM-2	1,000
MTM-3	3,000
MOST	12,000

motion time systems. A more detailed discussion of accuracy is presented in Appendix A, Theory.

Documentation

One of the most burdensome problems in the standards development process is the volume of paperwork required by the most widely used predetermined work measurement systems. MOST has shown that where the more detailed systems require between 40 and 100 pages of documentation, MOST requires as few as five. The substantially reduced amount of paperwork enables the analysts to complete studies faster and to update standards more easily. An example comparing the documentation required for four work measurement techniques is shown in Table 1.4. In this application, MOST produced not only a fraction of the documentation of the MTM techniques, but also a time value quite comparable to the more costly procedures. It is interesting to note that the reduction of paper generated by MOST does not lead to a lack of definition of the method used to perform the task. On the contrary, the method description found with the MOST Systems is a clear, concise, plain-language sentence describing the activity. The MOST method descriptions can and have been used for operator training.

Method Sensitivity

Too often, work study analysts perceive their jobs as simply providing the time required for an operation to take place. The result is that one of the analyst's most important functions, that of methods improvement, is often given little or no consideration. Especially vulnerable to this misconception is the time study analyst whose attention is necessarily focused on a watch. Also, when using time study, a quantitative comparison of methods cannot be produced unless another time study is taken of the new method. MOST, like any predetermined motion

Table 1.4 Comparison of Documentation Required

WORK MEASUREMENT TECHNIQUE	NO. OF DOCUMENTATION PAGES USED	ASSEMBLY OPERATION TIME, TMUs
MTM-1	16	4,402
MTM-2	10	4,445
MTM-3	8	4,950
MOST	1	4,530

Note: Actual numbers of pages produced may vary among analysts.

time system, is concerned primarily with the motions that make up an operation. The times or index values for these motions have already been predetermined and are immediately available to the analyst from data cards, or after experience, from memory. It is the analyst's responsibility to recognize the specific motion patterns and to assign the appropriate index values to each sequence model parameter. Since MOST index values are time related, they provide a quick means for evaluating the relative length of time required for performing a specific method. The analyst's attention is automatically focused on motions requiring longer times, such as subactivities with index values of 6 or greater. This is especially true since a complete MOST analysis will quite often require no more than one page. It is therefore easy for the analyst to rearrange the workplace layout or introduce a new tool or a fixture. By doing so, high index numbers may be reduced or even eliminated. The analyst can, on a copy of the analysis, change appropriate index numbers and thereby compute the savings resulting from the improved method.

MOST, then, is a method-sensitive technique; i.e., it is sensitive to the variations in time required by different methods. This feature is very effective in evaluating alternative methods of performing operations with regard to time and cost. The MOST analysis will clearly indicate the more economical and less fatiguing method.

The fact that MOST systems is method sensitive greatly increases its worth as a work measurement tool. Not only does it indicate the time needed to perform various activities, it also provides the analyst with an instant cue that a method should be reviewed. The results are clear, concise, easily understood time calculations that indicate the opportunities for saving time, money, and energy.

Applicability

For what situations can MOST be used? Because manual work normally includes some variation from one cycle to the next, MOST, with its statistically established time ranges and time values, can produce times comparable to more detailed systems for the majority of manual operations. Therefore, MOST is appropriate for any manual work that contains variation from one cycle to another. MOST should not be used in situations in which a cycle is *repeated identically* over a long period of time. In these situations, which, by the way, do not occur very often, a more detailed system should be chosen as the analytical tool. Additional information can be found in Appendix A, Theory.

2
The General Move Sequence

The General Move Sequence deals with the spatial displacement of an object. Under manual control, the object follows an unrestricted path through the air. If the object is in contact with, or restrained in any way by another object during the move, the General Move Sequence is not applicable.

Characteristically, General Move follows a fixed sequence of subactivities identified by the following steps.

1. Reach with one or two hands a distance to the object(s), either directly or in conjunction with body motions.
2. Gain manual control of the object.
3. Move the object a distance to the point of placement, either directly or in conjunction with body motions.
4. Place the object in a temporary or final position.
5. Return to workplace.

These five subactivities form the basis for the activity sequence describing the manual displacement of an object freely through space. This sequence describes the manual events that can occur when moving an object freely through the air and is therefore known as a sequence model. The major function of the sequence model is to guide the attention of the analyst through an operation, thereby adding the dimension of having a preprinted and standardized analysis format. The existence of the sequence model provides for increased analyst consistency and reduced subactivity omission.

The Sequence Model

The sequence model takes the form of a series of letters representing each of the various subactivities (called parameters) of the General Move Activity Sequence. With the exception of an additional parameter for body motions, the General Move Sequence is the same as the previous five-step pattern:

A B G A B P A

where: A = Action distance
B = Body motion
G = Gain control
P = Place

Parameter Definitions

A Action Distance

This parameter covers all spatial movement or actions of the fingers, hands, and/or feet, either loaded or unloaded. Any control of these actions by the surroundings requires the use of other parameters.

B Body Motion

This parameter refers to either vertical (up and down) motions of the body or the actions necessary to overcome an obstruction or impairment to body movement.

G Gain Control

This parameter covers all manual motions (mainly finger, hand, and foot) employed to obtain complete manual control of an object(s) and to subsequently relinquish that control. The G parameter can include one or several short-move motions whose objective is to gain full control of the object(s) before it is to be moved to another location.

P Place

This parameter refers to actions at the final stage of an object's displacement to align, orient, and/or engage the object with another object(s) before control of the object is relinquished.

Parameter Indexing

The displacement of an object through space occurs in three distinct phases, as shown by the following General Move Sequence breakdown.

GET / PUT / RETURN

A B G / A B P / A

The first phase, referred to as Get, describes the actions to reach to the object, with body motions (if necessary), and gain control of the object. The A parameter indicates the distance the hand or body travels in order to reach the object, while the B indicates the need for any body motions during this action. The degree of difficulty encountered in gaining control of the object is described by the G parameter.

The Put phase of the sequence model describes the actions to move the object to another location. As before, the A and B parameters indicate the distance the hand or body travels with the object and the need for any body motions during the move before the object is placed. The manner in which the object is placed is described by the P parameter.

The third phase simply indicates the distance traveled by the operator to Return to the workplace following the placement of the object.

The MOST analyst should strictly adhere to the three-phase breakdown of the General Move Sequence Model. Such adherence provides consistency in application and ease in communication. To acquire such consistency, the analyst should always ask these questions prior to indexing a sequence model:

What is the object being moved?
How is it moved (to determine the appropriate sequence model)?
What did the operator do to Get the object (to determine the index values for A, B, and G)?
What did the operator do to Put the object (to determine the index values for A, B, and P)?
Did the operator Return (to determine the index value for the final A)?

To ask these questions (or similar questions), when the Controlled Move Sequence or Tool Use Sequence Model is involved, ensures that measure of consistency vital to effective application of MOST.

Indexing each parameter of the General Move Sequence Model is accomplished by observing or visualizing the operator's actions during each phase of the activity and selecting the appropriate parameter variants from the data card (Table 2.1) that describes those actions. The corresponding index value for each parameter is taken from the extreme left or right column of the data card and is written just below and to the right of the sequence model parameter, for example, A_3.

Consider the example of a machine operator getting a finished part from a workplace table and putting it on a pallet. Assume that the operator is standing directly in front of the part, which is light in weight; the location of the pallet is 10 steps away on the floor. The sequence model for this activity is filled out as follows.

A_1 B_0 G_1 A_{16} B_6 P_1 A_{16}

Since the operator is standing directly in front of the part, the first A parameter in the sequence is indexed A_1 because the part is located within reach. (Look under the action distance column of the data card (Table 2.1) for Within Reach and note the corresponding index value to the left.) No Body Motion is needed to reach the part; therefore, a 0 is assigned to the Body Motion parameter (B_0), while control of the object is gained with no difficulty–G_1 (light object under Gain Control column). The part is then moved 10 steps away (A_{16}) and placed on the floor, B_6 (Bend and Arise). No difficulty is encountered in placing the part on a pallet; it is simply Laid Aside, P_1. The operator then walks back (Returns) to the workplace, which is 10 steps away (A_{16}).

The time to perform this activity is computed by simply adding all index numbers in the sequence and multiplying by 10 to convert to TMU: $(1 + 0 + 1 + 16 + 6 + 1 + 16) \times 10 = 410$ TMU.

In the remaining sections of this chapter, the parameter variants for each of the General Move Sequence parameters are examined in detail. The parameter values up to and including index value 16 (i.e., all values on the General Move data card) should be familiar enough to the MOST analyst to be applied from memory. Because of this, the majority of work performed within the confines of a well-designed workplace can be analyzed without the aid of the data card.

Action Distance (A)

Action Distance covers all spatial movement or actions of the fingers, hands, and/or feet, either loaded or unloaded. Any control of these actions by the surroundings requires the use of other parameters.

$A_0 \leqslant 2$ in. (5 cm)

Any displacement of the fingers, hands, and/or feet a distance less than or equal to 2 in. (5 cm) will carry a zero index value. Time for performing these short distances are included within the Gain Control and Place parameters. *Example:* Used when reaching between the number keys on a pocket calculator or placing nuts or washers on bolts located less than 2 in. (5 cm) apart.

A_1 Within Reach

Actions are confined to an area described by the arc of the outstretched arm pivoted about the shoulder. With body assistance–a short bending or turning of the body from the waist–this "within reach" area is extended somewhat. However, taking a step for further extension of the area exceeds the limits of an A_1

Table 2.1 General Move Data Card

m	ABGABPA	GENERAL MOVE			
INDEX	A	B	G	P	INDEX
	ACTION DISTANCE	BODY MOTION	GAIN CONTROL	PLACE	
0	≤ 2 IN ≤ 5 CM			HOLD TOSS	0
1	WITHIN REACH		LIGHT OBJECT LIGHT OBJECTS SIMO	LAY ASIDE LOOSE FIT	1
3	1-2 STEPS	BEND AND ARISE 50% OCC	NON SIMO HEAVY OR BULKY BLIND OR OBSTRUCTED DISENGAGE INTERLOCKED COLLECT	ADJUSTMENTS LIGHT PRESSURE DOUBLE	3
6	3-4 STEPS	BEND AND ARISE		CARE OR PRECISION HEAVY PRESSURE BLIND OR OBSTRUCTED INTERMEDIATE MOVES	6
10	5-7 STEPS	SIT OR STAND			10
16	8-10 STEPS	THROUGH DOOR CLIMB ON OR OFF			16

and must be analyzed with an A_3 (one to two steps). *Example:* In a well laid out workstation such as that shown in Figure 2.1, all parts and tools can be reached without displacing the body by taking a step.

The parameter value A_1 also applies to the actions of the leg or foot reaching to an object, lever, or pedal. If the trunk of the body is shifted, however, the action must be considered a step (A_3).

A_3 One to Two Steps

The trunk of the body is shifted or displaced by walking, stepping to the side, or turning the body around using one or two steps. Steps refers to the total number of times each foot hits the floor.

Index values for longer-action distances involving walking are found in Table 2.2. While these values generally refer to the horizontal movement of the body, they also apply to walking up or down normally inclined stair steps. Index values are given in terms of both Steps and Distance. When using this table, however, the preferred method is to count the number of steps taken. The reason is

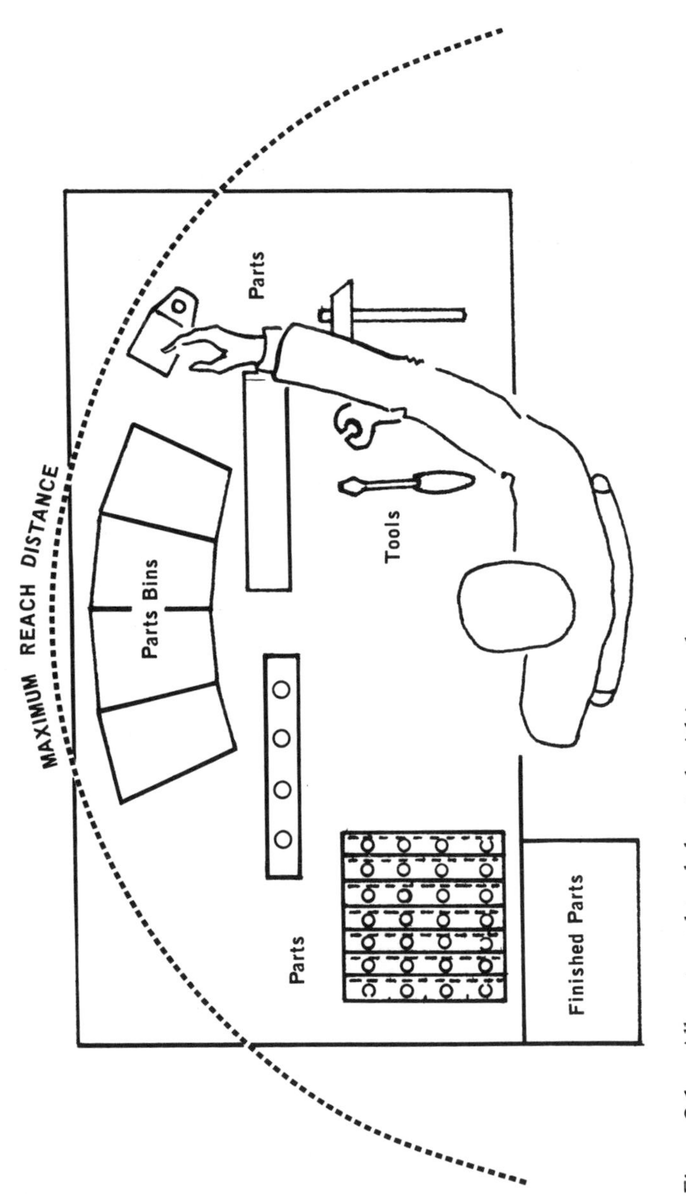

Figure 2.1 All parts and tools located within reach.

Table 2.2 Extended Action Distance

ACTION DISTANCE			
INDEX	STEPS	DISTANCE (FT)	DISTANCE (M)
24	11-15	38	12
32	16-20	50	15
42	21-26	65	20
54	27-33	83	25
67	34-40	100	30
81	41-49	123	38
96	50-57	143	44
113	58-67	168	51
131	68-78	195	59
152	79-90	225	69
173	91-102	255	78
196	103-115	288	88
220	116-128	320	98
245	129-142	355	108
270	143-158	395	120
300	159-174	435	133
330	175-191	478	146

Note: The values in the columns Distance (Ft.) and (M) are read up to and including.

that research has shown the time required to take a step to be relatively constant regardless of the size of the load carried. In other words, the time required to take five steps while carrying a heavy load is the same as the time required to take five steps with no load. However, the influence of the load is to shorten the step length, thereby increasing the number of steps required to cover a specific distance. In this way, the effect of any load is reflected in the action distance parameter. Therefore, whenever possible, action distance values should be based on the number of steps taken by the operator not the distance traveled.

Occasionally, however, it is not possible to observe the operator at work. If this is the case, action distance values must be determined from distances measured at the workplace or obtained from drawings. The distances in Table 2.2 are based on an average step length of 2½ ft. (0.75 m). *Note:* the action distance values were generated to include walking in a normal manufacturing environment, and as a result they include an average pace of 2½ ft. (0.75 m), obstructed and unobstructed walking, walking up or down normally inclined stairs, and walking with or without weight. Should a particular job contain a high quantity of long (those exceeding 478 ft., 146 m), unobstructed, and unencumbered walking distances, the action distances provided may not be appropriate, and the values should then be validated.

Body Motion (B)

Body Motion refers to either vertical (up and down) motions of the body or the actions necessary to overcome an obstruction or impairment to body movement.

B_6 Bend and Arise

From an erect standing position, the trunk of the body is lowered by bending from the waist and/or knees to allow the hands to reach below the knees and subsequently return to an upright position. It is not necessary, however, for the hands to actually reach below the knees, only that the body be lowered sufficiently to allow the reach. B_6 may be simply bending from the waist with the knees stiff, stooping down by bending at the knees, or kneeling down on one knee. See Figure 2.2.

B_3 Bend and Arise, 50% Occurrence

Bend and Arise is required only 50% of the time during a repetitive activity, such as stacking or unstacking several objects. In stacking, (Figure 2.3), the first few objects may require a full Bend and Arise to place the objects at floor level. As the stack becomes taller, the last objects for stacking require no body motions at all.

Figure 2.2 Examples of bend and arise. Notice that in each case the hands are able to reach below the knees.

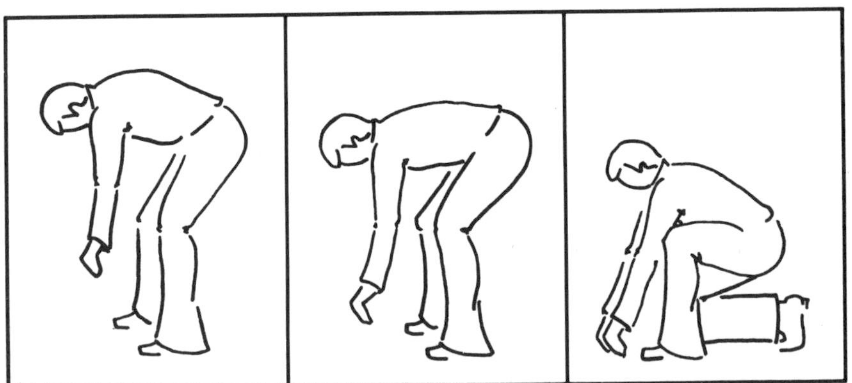

B_{10} Sit or Stand

When the act of sitting down or standing up requires a series of several hand, feet, and body motions to move a chair or stool into a position that allows the body to either sit or stand, a B_{10} is appropriate. All the motions to manipulate the chair and body are included in the B_{10} body motion. If the chair or stool is stationary, and several feet and body motions are necessary to either situate the body comfortably in the seat or to climb on or off the stool, a B_{10} would also apply. Note that B_{10} covers either sit *or* stand, not both.

A special situation may be encountered in industry where an operator sits or stands without moving the chair, such as sitting on a bench. A special index value B_3, Sit or Stand without Moving Chair, could be used in that situation.

B_{16} Through Door

Passing through a door normally consists of reaching for and turning the handle, opening the door, walking through the doorway, and subsequently closing the door. This parameter variant covers any hinged, double, or swinging door.

Note that the three or four steps required to walk through the doorway are included within the B_{16} value. These steps should not be added to the action distance parameter or subtracted from it. *Example:* An operator walks five steps to a door, passes through the door, and walks three steps to a desk where a light object is picked up and placed on the floor beside the desk.

Figure 2.3 Bend and arise, 50% occurrence.

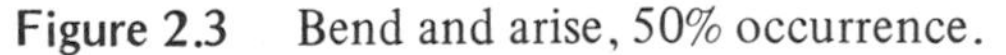

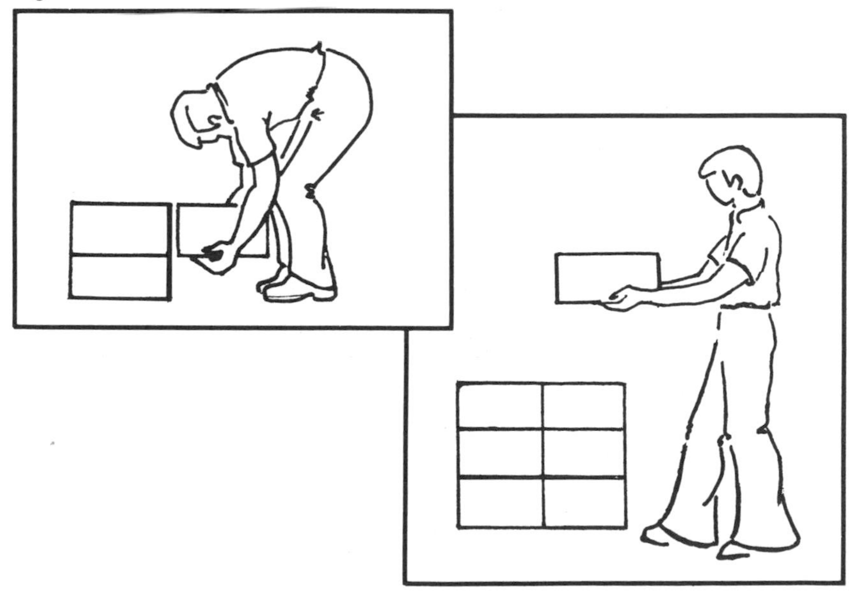

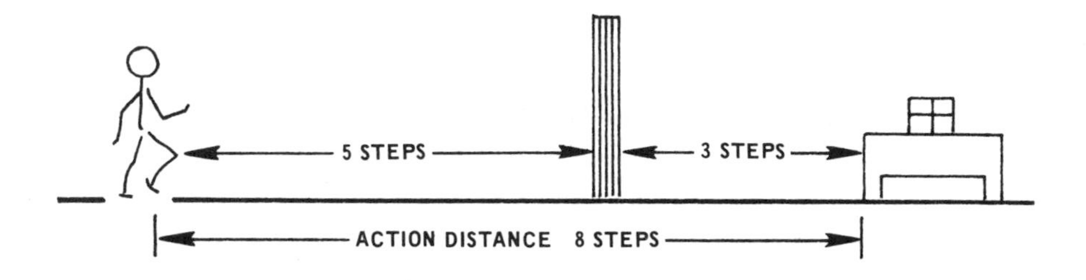

The action distance to Get the object is eight steps (A_{16}). The steps to actually pass through the doorway are included in the value B_{16}. The appropriate analysis would be:

GET			PUT			RETURN	
A_{16}	B_{16}	G_1	A_1	B_6	P_1	A_0	410 TMU

B_{16} Climb On or Off

This parameter variant covers climbing on or off a work platform or any raised surface (approximately 3 ft. or 1 m high), using a series of hand and body motions to lift or lower the body. Climbing onto a platform is accomplished by first placing one hand on the edge and then lifting the knee to the platform. By placing the other hand on the platform and bending forward, the weight of the body is then shifted, allowing the other knee to be lifted onto the platform. The activity is completed by arising from both knees. Climbing off the platform consists of the same actions, only performed in the reverse order. *Example:* Climb onto a truck frame on an assembly line to attach a bracket for the exhaust system.

Gain Control (G)

Gain Control covers all manual motions (mainly finger, hand, and foot) employed to obtain complete manual control of an object and to subsequently relinquish that control. The G parameter can include one or several short-move motions whose objective is to gain full control of the object(s) before it is to be moved to another location.

G_1 Light Object

Any type of grasp can be used as long as no difficulty is encountered as described by the G_3 parameter variants. The object may be jumbled with other objects, lying close against a flat surface, or simply lying by itself. Control may be gained by simply touching the object with the fingers, hand, or foot (contact grasp), or it may require a more difficult grasping action such as that needed to pick one object out of a jumbled pile of objects. Either one or two hands may be used as long as only one object is obtained and that object is accessible for the simultaneous grasps of both hands. If several objects are grouped together or arranged in such a way that they may be picked up as one object, G_1 will still apply.

Examples: Pick up a hammer from a work bench. Obtain one washer from a parts bin full of washers. Using both hands, pick up a manual lying by itself. Obtain one sheet of paper from the top of a desk. Pick up pencils grouped to-together in a holder (several objects grouped as one). Obtain one bolt from a jumbled pile of bolts. Grasp a lever, crank, knob, toggle switch, pushbutton, foot pedal, or other activating device (to be applied in the Controlled Move Sequence Model).

G_1 Light Objects Simo

Simo refers to manual actions performed simultaneously by different body members. That is, one hand gains control of a light object (G_1), while at the same time the other hand obtains another light object (G_1). The total time, then, is no more than that required to gain control of one light object. *Examples:* Using both hands, pick up a hammer and nail lying side by side. Using both hands, pick up two small suitcases. Pick up a pencil and a straightedge using both hands.

G_3 Non Simo

Due to the nature of the job or the conditions under which the job is performed, the operator is unable to gain control of two objects or of two suitable grasping points of one object simultaneously. While one hand is grasping an object, the other hand must wait before it can grasp the other object. Therefore, gain control time must be allowed for both hands, hence the larger index value G_3 applies.

The ability of the operator to perform simultaneous motions is largely dependent on the amount of practice opportunity available. For example, an assembly operator who continuously gets parts from the same two locations will have no trouble performing the activity simo. After repeating a number of cycles, the operator develops an automatic reaction to the exact location of each part.

On the other hand, simultaneous motions will sometimes be difficult for workers in a job shop. Because of the infrequent occurrence of many operations, the operator will have little practice opportunity to gain the automatic skills necessary to perform simo motions.

Regarding selection of the Simo versus Non Simo parameter, the analyst should observe the operator's actions wherever possible. Normally, simo actions can easily be recognized by their automatic appearance. (For further discussion, see Method Levels in Chapter 7).

G_3 Heavy or Bulky

Control of heavy or bulky objects is achieved only after the muscles are tensed to a point where the effects of the difficulty created by the weight, shape, or size are overcome. We can identify this variant by the *hesitation or pause* needed for the attainment of sufficient muscular force required to move the object.

This effect is influenced not only by the actual weight of the object, but also by the location of the object with respect to the body, the existence of handles or grips for easy grasping, or even the strength of the individual. Poorly located objects, even smaller or lighter ones, for example, may require some hesitation or the movement of the body for balance or additional muscular control for leverage. With the existence of handles or other easy grasping devices located appropriately on the object, the effect of the weight can be significantly reduced.

Therefore, when considering this parameter variant for gain control, the major criterion is not the actual weight of the object, but the hesitation or pause needed for the muscles to tense or the body to stiffen prior to moving the object. See Figure 2.4.

Examples: Get hold of an automobile battery located on the floor. Take hold of a loaded hand cart before pulling. Get hold of a briefcase by reaching over other baggage. Brace your body before pushing a heavy carton across a bench. Pick up a large, empty television packing box.

The weight or bulk of an object can also affect the method of gaining control. Before a heavy or bulky object can be completely controlled, it may be necessary to move or reorient the object. This might require obtaining a temporary grip and sliding the object closer to the body before complete control of the

Figure 2.4 Examples of G_3, gain control of heavy or bulky objects.

object is obtained (see Figure 2.5). In extreme cases where several "intermediate moves" of the object are required, analysis is accomplished through the use of additional parameters or sequence models if necessary. For example, use a Controlled Move Sequence Model to analyze sliding the object closer. However, we suggest that the method in such a case be reviewed and improved if possible.

G3 Blind or Obstructed

The accessibility to the object is restricted because an obstacle either prevents the operator from seeing the object or creates an obstruction to the hand or fingers when attempting to gain control of the object. If the location is blind, the operator must *feel around* for the object before it can be grasped. When an obstruction presents itself, the fingers or hand must be *worked around* the obstacle before reaching the object.

Examples: Obtain a washer from a stud located on the other side of a panel (blind). Work the fingers around the wiring in an electrical assembly to get a part (obstructed). Get a pocket knife from your pocket.

G3 Disengage

The application of muscular force is needed to free the object from its surroundings. Disengage is characterized by the application of pressure (to overcome resis-

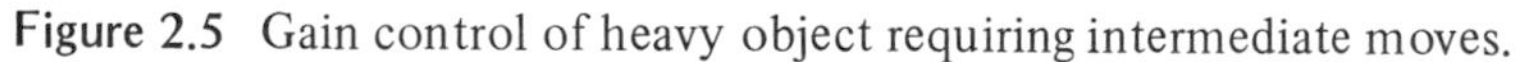

Figure 2.5 Gain control of heavy object requiring intermediate moves.

tance), followed by the sudden movement and recoil of the object. The recoil of the object, however, must follow an unrestricted path through the air (not to be confused with unseating a lever, crank, or other controlled devices. *Examples:* Remove a tightly fitting socket from a rachet tool. Remove a knife stuck in wood. Remove the cork from a wine bottle.

G_3 Interlocked

The object is intermingled or tangled with other objects and must be separated or worked free before complete control is achieved. *Examples:* Remove a hammer from a crowded tool box (the hammerhead is buried beneath other tools). From a box of springs, gain control of one spring that is tangled with another.

G_3 Collect

Control of several objects is accomplished. The objects may be jumbled together in a pile or spread out over a surface. If jumbled, control of several objects is achieved by digging down into the pile with the hand(s) and bringing up a handful. When spread out, the objects may be swept together with the hand(s) and fingers and picked up as one object. If the items are in very close proximity to one another and picked up individually, G_3 will cover gaining control of up to two objects (regrasps) per hand (i.e., a total of four if two hands are used simultaneously). If even more items are grasped individually, a series of G_1 or additional G_3 grasping actions may be observed. Repeated or additional reaches (A_1) may be included if required. *Note:* Ensure that the best method is being performed and these additional grasps, etc., are in fact necessary.

Examples: Grasp a handful of nails from a bin. Collect several sheets of paper lying on a desk. Get a handful of change from your pocket. Gather up a pen, pencil, and eraser spread out on a desk with one sweeping motion of the hand. Individually collect two bolts lying close together from the top of a workbench.

Place (P)

Place refers to actions occurring at the final stage of an object's displacement to align, orient, or engage the object with another before control of the object is relinquished. Basically, the index value for the place parameter is chosen by the amount of difficulty encountered in placing the object.

P_0 Hold

The placing parameter does not occur. The object is picked up and held. Placement occurs in a later sequence model. *Examples:* Pick up a book. Pick up a job card.

P_0 Toss

The placing parameter does not occur. The object is released during the previous move (action distance parameter) without any placing motions or pause to point the object toward the target. *Examples*: Toss a finished part into a tote bin. Toss a completed assembly down a drop chute. Toss balled-up paper into a trash can.

P_1 Lay Aside

The object is simply placed in an approximate location with no apparent aligning or adjusting motions. This placement requires low control by the mental, visual, or muscular senses. *Examples:* Lay a hand tool aside after using. Place a pencil on a desk. Lay a manual on a table.

P_1 Loose Fit

The object is placed in a more specific location than that described by the Lay Aside parameter, but tolerances are such that only a very modest amount of mental, visual, or muscular control is necessary to place it. The clearance between the engaging parts is loose enough so that a single adjusting motion, without the application of pressure, is required to seat or position the object.

Examples: Place a washer on a bolt. Replace a telephone receiver on the hook. Place a coat hanger on a rack. Place a dull pencil in a sharpener.

P_3 Adjustments

Adjustments are defined as the unintentional actions occurring at the point of placement caused by difficulty in handling the object, closeness of fit, lack of symmetry of the engaging parts, or uncomfortable working conditions. These adjustments are recognized as obvious fumbles, hesitations, or correcting motions at the point of placement to align, orient, and/or engage the object.

Examples: Place a key in a lock. Align a center punch to scribe a mark. Place a screw on a threaded junction and pick up the threads.* Place the looped end of a small piece of wire around a terminal.

P_3 Light Pressure

Due to close tolerances or the nature of the placement, the application of muscular force is needed to seat the object even if the initial positioning action could be classified as loose (P_1). This could occur, for example, as the snapping action required to seat a socket on a ratchet tool. *Examples:* Place a moist stamp on an

*Threaded positions are nearly always a P_3, unless they are either blind or obstructed (P_6), or placed in a deep hole (P_1), where the threaded pick-up action is not required.

envelope. Press a thumb tack into a cork board. Insert an electric plug into a socket (light muscular force is required to seat the object after orienting with a single adjustment).

P_3 Double

Two distinct placements occur during the total placing activity. For example, to assemble two parts held by a fixture, a bolt is first placed through a hole in both parts before the nut is placed on the bolt with the other hand (Figure 2.6). The first placement occurs with the bolt through the holes followed by a second placement of the nut on the bolt.

This parameter can also be applied to an object being lined up to two different marks following a general move. For P_3 to apply, however, these marks must be within 4 in. (10 cm) of each other. If more than 4 in. (10 cm) exist between each mark, special eye times are needed which require additional care in the placement (P_6). (For more detailed information, see Align in the Controlled Move Sequence.)

Example: Place an original between the two marks on a photocopy machine.

P_6 Care or Precision

Extreme care is needed to place an object within a closely defined relationship with another object. The occurrence of this variant is characterized by the obvious slow motion of the placement due to the high degree of concentration required for mental, visual, and muscular coordination. *Examples:* Thread a needle. Place a soldering iron to a crowded circuit connection.

P_6 Heavy Pressure

As a result of very tight tolerances, not the weight of an object alone, a high degree of muscular force is needed to engage the object. Occurring only rarely in practice, heavy pressure can be easily recognized as the regrasping of the object,

Figure 2.6 Left: Place bolt through hole before placing nut. Right: Fasten one object to another.

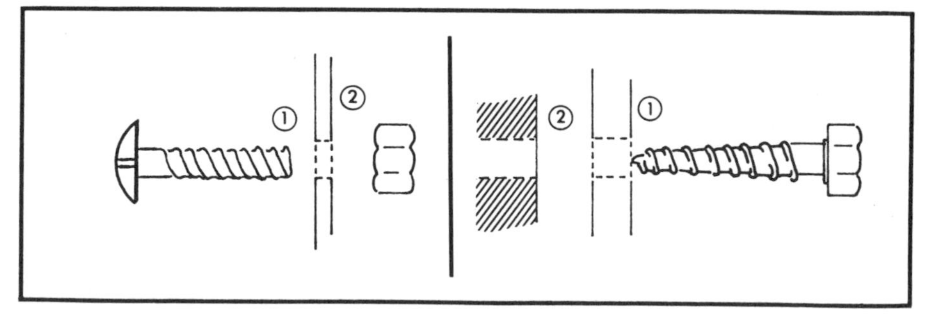

tensing of the muscles, and the preparation of the body prior to the application of pressure. *Example:* Place a book in a very tight slot on a bookshelf.

Note: An index value for P is never chosen by the weight of the object alone. While weight may influence the difficulty in placement, it is that difficulty which determines the value chosen for P, not the weight. For example, a heavy suitcase might simply be laid to rest on the floor, in which case a P_1 (lay aside) would be chosen, while a light package might be squeezed into a tight space between two other boxes on a shelf; there, a P_6 (heavy pressure) might be appropriate.

P_6 Blind or Obstructed

Conditions are similar to those encountered by the Gain Control parameter with the same title. The accessibility to the point of placement is restricted because an obstacle either prevents the operator from seeing the point of placement or creates an obstruction to the hand or fingers when attempting to place the object. If the location is blind, the operator must *feel around* for the placement location before the object can actually be placed with adjustments. When an obstruction presents itself, the fingers and/or hands must be *worked around* the obstacle before placing the object with adjustments.

Examples: Place a nut on a hidden bolt. Place a spark plug in an engine block after working the hands in between the distributor wiring.

P_6 Intermediate Moves

Several intermediate moves of the object are required before placing it in a final location. These intermediate moves are necessary because the nature of the object or the conditions surrounding the object prevent direct placement. With heavy, bulky, or difficult-to-handle objects, this parameter is recognized as a series of placing, shifting of grasps, and moving actions occurring before final placement. This additional handling is needed to overcome the awkward nature of the object.

Examples: Place chairs in a neat row by first setting a chair down and then aligning it with several sliding moves. Place a large box down on its corner and "walk it" into position. Place a heavy or bulky box on a pallet and stack neatly.

A special case of this variant is encountered when placing one object from a handful of different objects in the palm. Before actually placing the object, several finger and hand movements are required to select and shift one of the objects from the palm to the fingertips. This unpalming action is more than a simple regrasp. The hand must first be turned over, allowing visual selection of the appropriate object. Several complex finger motions (intermediate moves) are then needed to shift the object up to the fingertips before placement can occur.

Note: This case (P_6) applies only to a handful of different objects. If the objects held in the palm are all similar, visual selection is not necessary. Therefore, a simple regrasp will be sufficient for unpalming any of the objects. As this regrasp normally occurs during the action distance to place the object, no additional regrasp time is needed. The value for P is then chosen from the data card by the amount of difficulty required to place the object.

Examples of the special case P_6 intermediate moves: From a handful of change, use the thumb to push a dime to the fingertips and place it in a vending machine. Using the thumb, select a ½-inch washer from a handful of assorted washers and nuts and place it on a bolt.

Special Parameter Variants

The parameter variants found on the data card and defined above cover most activities observed in normal production situations. However, special situations are always a possibility and must also be analyzed. Two such cases that may appear in work observed by the MOST analyst are defined below. Their infrequent occurrence does not merit their placement on the General Move data card; however, the need for a consistent analysis approach and consistent definitions is considered necessary.

B_3 Sit or Stand without Moving Chair

When the body is simply lowered into a chair from an erect position, with no hand or foot motions required to manipulate the chair, or it is raised from a seated position without the aid of any hand or foot motions, then sit or stand without moving chair (B_3) is appropriate. Note that this value covers either sit or stand, not both. *Examples:* Lower the body to a sitting position on a bench. Stand up from a theater seat.

P_3 Loose Fit Blind

Conditions are similar to those encountered by the Gain Control parameter with the same title. In such a situation, the operator must feel around for the placement location before a loose placement can occur. *Examples:* Place a washer on a hidden stud. Place a cotter pin through a hole located behind a panel.

Placement with Insertion

Occasionally, following initial placement, an object may be inserted to reach its destination. This insertion, while technically a controlled move (Chap. 3), can be indexed on the final A of the General Move Sequence Model. This convention was established to simplify the analysis procedure; otherwise, an additional Controlled Move Sequence Model would be needed to describe only one parameter (M). As a general rule when using the final A, an A_1 should follow the P parameter to represent the insertion of an object from 2 in. (5 cm) up to and including 12 in. (30 cm). *Example:* Replacing the oil dipstick in the engine block of an automobile would be indexed:

$A_0 \quad B_0 \quad G_0 \quad A_1 \quad B_6 \quad P_3 \quad A_1$

In addition to an insertion of from 2 to 12 in. (5-30 cm) the last A of the general move sequence model is normally used for:

- Return to original location (walking only)
- Retracting the hand from inside a machine

Parameter Frequencies

Often, one or more parameters within the General Move Sequence may occur more than once, for example, when placing several objects from a handful. This activity is shown on the sequence model by placing parentheses around the parameters which are repeated and writing the number of occurrences in the frequency column of the calculation sheet (see Chap. 7), also within parentheses. The time calculation is performed as follows.

1. Add all index values for the parameters within parentheses.
2. Multiply this value by the number of occurrences (the number within parentheses in the frequency column).
3. Add this product to the remaining parameter index values.
4. Convert the total to TMU by multiplying by 10.

If an entire sequence occurs more than once, the number of occurrences is placed in the frequency column without parentheses. The time calculation is performed by taking the total TMU for the sequence model times the frequency. *Example:* Get a handful of washers and place on six bolts located 5 in. (12 cm) apart.

$A_1 \quad B_0 \quad G_3 \quad (A_1 \quad B_0 \quad P_1) \quad A_0 \quad (6)$

GET	A_1	Reach to washers
	B_0	No body motion
	G_3	Collect a handful of washers
PUT	A_1	Reach to place a washer
	B_0	No body motion
	P_1	Place washer, loose fit
RETURN	A_0	No return

As indicated, only the parameters in the Put section of this sequence model are repeated 6 times. The operator reaches A_1 with no body motions B_0 and places a washer P_1.

The time calculation steps are:

1. $(A_1 B_0 P_1) = (1 + 0 + 1) = 2$
2. $2 \times 6 = 12$
3. $1 + 0 + 3 + 12 + 0 = 16$
4. $16 \times 10 = 160$ TMU

These four steps could also be written:

$[(1+1) \times (6)+1+3] \times 10 = 160$ TMU

Additionally, if the entire sequence (the placement of six washers) occurs twice, the following analysis would apply.

$A_1 \quad B_0 \quad G_3 \quad (A_1 \quad B_0 \quad P_1) \quad A_0 \quad (6) \quad 2$

$[(1+0+1) \times (6)+1+0+3+0] \times 10 \times 2 = 320$ TMU

The above situation, in which the Put section of the sequence model is repeated, illustrates only one of the countless situations involving frequencies which could occur. Actually, any one or any combination of parameters could have a frequency applied to them. The frequency can be a whole number, decimal, or fraction. *Note:* Only one set of parentheses is allowed per sequence model.

General Move Examples

1. A man walks four steps to a small suitcase, picks it up from the floor, and without moving further places it on a table located within reach.

$A_6 \quad B_6 \quad G_1 \quad A_1 \quad B_0 \quad P_1 \quad A_0$

$(6 + 6 + 1 + 1 + 1) \times 10 = 150$ TMU

2. An operator standing in front of a lathe walks six steps to a heavy part lying on the floor, picks up the part, walks six steps back to the machine, and places it in a 3-jaw chuck with several adjusting actions. The part must be inserted 4 in. (10 cm) into the chuck jaws.

$A_{10} \quad B_6 \quad G_3 \quad A_{10} \quad B_0 \quad P_3 \quad A_1$

$(10+6+3+10+3+1) \times 10 = 330$ TMU

3. From a stack located 10 ft. (3 m) away, a heavy object must be picked up and moved 5 ft. (2 m) and placed on top of a workbench with some adjustments. The height of this stack will vary from waist to floor level. Following the placement of the object on the workbench, the operator returns to the original location, which is 11 ft. (3.5 m) away.

A_6 B_3 G_3 A_3 B_0 P_3 A_{10}

$(6 + 3 + 3 + 3 + 3 + 10) \times 10 = 280$ TMU

4. An assembly worker gets a handful of washers (six) from a bin located within reach and places one on each of six bolts located within reach which are 4 in. (10 cm) apart.

A_1 B_0 G_3 $(A_1$ B_0 $P_1)$ A_0 (6)

$[(1 + 0 + 1) \times (6) + 1 + 0 + 3 + 0] \times 10 = 160$ TMU

3
The Controlled Move Sequence

The Controlled Move Sequence describes the manual displacement of an object over a "controlled" path. That is, movement of the object is restricted in at least one direction by contact with or an attachment to another object.

Similar to the General Move Sequence, Controlled Move follows a fixed sequence of subactivities identified by the following steps.

1. Reach with one or two hands a distance to the object, either directly or in conjunction with body motions.
2. Gain manual control of the object.
3. Move the object over a controlled path.
4. Allow time for a process to occur.
5. Align the object following the controlled move or at the conclusion of the process time.
6. Return to workplace.

These six subactivities form the basis for the activity sequence describing the *manual displacement* of an object over a *controlled path.*

The Sequence Model

The sequence model takes the form of a series of letters representing each of the various subactivities (called parameters) of the Controlled Move Activity Sequence.

A B G M X I A

where A = Action distance
B = Body motion

G = Gain control
M = Move controlled
X = Process time
I = Align

Parameter Definitions

Only three new parameters are introduced, as the A, B, and G parameters were discussed with the General Move Sequence and remain unchanged.

M Move Controlled

This parameter covers all manually guided movements or actions of an object over a controlled path.

X Process Time

This parameter occurs as that portion of work controlled by processes or machines and not by manual actions.

I Align

This parameter refers to manual actions following the controlled move or at the conclusion of process time to achieve the alignment of objects.

Parameter Indexing

A Controlled Move is performed under one of two conditions. The object or device is restrained by its attachment to another object, such as a push button, lever, door, or crank, or it is controlled during the move by the contact it makes with the surface of another object, such as pushing a box across a table. In either case, *the object is not moved spatially* from one location to another. Otherwise, if neither condition affects the move, it must be considered a General Move.

A breakdown of the Controlled Move Sequence Model reveals that, like the General Move, three phases occur during the Controlled Move activity.

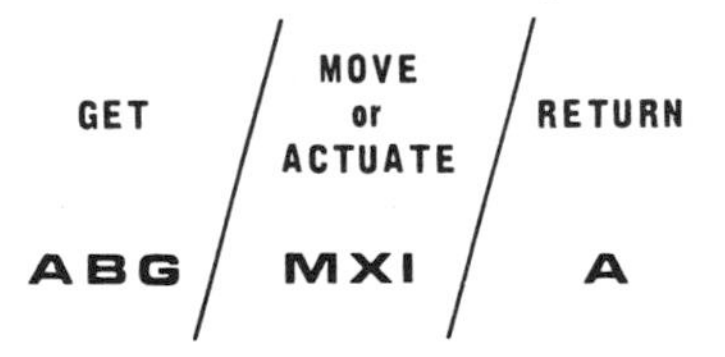

The Get and Return phases of Controlled Move carry the same parameters found in the General Move Sequence Model and, therefore describe the same subactivities. The fundamental difference is the activity immediately following the G parameter. This phase describes actions either to simply Move an object over a controlled path or to Actuate a control device. Normally, Move implies that the M and I parameters of the sequence model are involved, while Actuate usually applies to situations involving the M and X parameters. Of course, for either situation (Move or Actuate) any or all the parameters in the sequence model could be used, and all should be considered. A Move, for example, would occur when opening a tool cabinet door or sliding a box across a table. Engaging the clutch on a machine or flipping an electrical switch to start a process are examples of Actuate.

Like General Move, parameters in the Controlled Move Sequence Model are indexed by referring to a data card. Since the A, B, and G parameters are found on the General Move data card (Table 2.1), the Controlled Move data card (Table 3.1) includes only the M, X, and I parameters.

Parameter indexing is accomplished by selecting from the data card, the parameter variant that appropriately describes the observed or visualized Controlled Move and then applying the corresponding index value to the sequence model.

Table 3.1 The Controlled Move Sequence Data Card*

INDEX	M		X			I	INDEX
	ABGMXIA CONTROLLED MOVE SEQUENCE						
	MOVE CONTROLLED		PROCESS TIME			ALIGN	
	PUSH/PULL/PIVOT	CRANK (REVS.)	SECONDS	MINUTES	HOURS	OBJECT	
1	≤ 12 INCHES (30 CM) BUTTON/SWITCH/KNOB	-	.5	.01	.0001	TO ONE POINT	1
3	> 12 INCHES (30 CM) RESISTANCE, SEAT OR UNSEAT HIGH CONTROL 2 STAGES ≤ 12 INCHES (30 CM)	1	1.5	.02	.0004	TO TWO POINTS ≤ 4 INCHES (10 CM)	3
6	2 STAGES > 12 INCHES (30 CM)	3	2.5	.04	.0007	TO TWO POINTS > 4 INCHES (10 CM)	6
10	3 - 4 STAGES	6	4.5	.07	.0012		10
16		11	7.0	.10	.0019	PRECISION	16

*Values are read "up to and including."

Move Controlled (M)

Move Controlled covers all manually guided movements or actions of an object(s) over a controlled path. Index values for the M parameter are tabulated under two separate categories on the Controlled Move data card. The most frequently occurring parameter variants of Move Controlled (M) fall under the general heading Push/Pull/Pivot. The "crank" category applies to a special type of Controlled Move dealing with cranks, handwheels, or other devices requiring a circular "cranking" motion.

The following parameter variants apply to moves of an object or device that is hinged or pivoted at some point (e.g., a door, lever or knob) or restricted due to its surroundings (e.g., by guides, slots, or friction from surface).

$M_1 \leqslant 12$ in. (30 cm)

Object displacement is achieved by a movement of the fingers, hands, or feet not exceeding 12 in. (30 cm). *Examples:* Engage the feed on a cutting machine with a short hand lever. Press a light clutch pedal with the foot. Open a hinged lid on a small tool box. Push a box 10 in. (25 cm) across a workbench.

M_1 Button/Switch/Knob

The device is actuated by a short pressing, moving, or rotating action of the fingers, hands, wrist, or feet. *Examples:* Press a telephone hold button. Flip a wall light switch. Turn a door knob.

M_3 > 12 in. (30 cm)

Object displacement is achieved by a movement of the hands, arms, or feet exceeding 12 in. (30 cm). The maximum displacement covered by this parameter occurs with the extension of the arm plus body assistance. *Examples:* Push a carton across conveyor rollers. Pull a chain hoist full length. Close a cabinet door by pulling it shut. Open a file drawer full length.

In special situations, an object is pushed with a Controlled Move and then the operator will walk along with the object. For example, a warehouse worker reaches to a heavy box and pushes it along flat conveyor rollers a distance of 25 ft. (7.5 m). To analyze such situations, a special convention is used. A Controlled Move Sequence Model is chosen (for the box is moved over a controlled path), and the initial motion of the box is analyzed with the A, B, G, M, X, and I parameters. The Action Distance the operator walks with the box, plus any distance to Return can be analyzed by using the final A of the sequence model. The analysis for the above mentioned example is:

A_1 B_0 G_1 M_3 X_0 I_0 A_{16} 210 TMU

$(1 + 1 + 3 + 16) \times 10 = 210$

M_3 Resistance, Seat/Unseat

Conditions surrounding the object or device require that a resistance be overcome either prior to, during, or following the Controlled Move. This parameter variant covers the muscular force applied to "seat" or "unseat" an object or, if necessary, the short manual actions employed to latch or unlatch the object. Also, the object is moved and a resistance is present throughout the move. *Examples:* Engage the emergency brake on an automobile. Twist on a radiator cap securely. Push a heavy box across a table.

M_3 High Control

Care is needed to maintain or establish a specific orientation or alignment of the object during the Controlled Move. Characterized by a higher degree of visual concentration, this parameter variant is sometimes recognized by noticeably slower movements to keep within tolerance requirements or prevent injury or damage. The successful performance of this Controlled Move demands that eye contact be made with the object and its surroundings either during or at the end of the move. *Examples:* Turn the dial on a combination lock to a specific number. Slide a wood block on a machine table to position a mark on the block to a bandsaw blade prior to cutting. Slide a fragile item carefully across a workbench.

M_3 Two Stages $\leq$ 12 in. (30 cm)

An object is displaced in two directions or increments a distance not exceeding 12 in. (30 cm) per stage *without relinquishing control. Examples:* Engage and subsequently disengage the feed on a cutting machine with a short hand lever. Open and subsequently close a small tool box. Shift from the first to the third gear with a manual gearshift.

M_6 Two Stages $>$ 12 in. (30 cm)

An object is displaced in two directions or increments a distance exceeding 12 in. (30 cm) per stage without relinquishing control. *Examples:* Open and subsequently close a cabinet door. Shift a lever back and forth more than 12 in. (30 cm) in each direction. Raise and lower the cover of a copying machine.

M_{10} Three to Four Stages

An object is displaced in three or four directions or increments without relinquishing control. *Examples:* From the reverse position, shift to the first gear on a 4-speed automobile transmission. Set the feed/speed selector on an engine lathe.

Note: For those situations where a two-stage move exists, but the distance the object is moved over one stage is $\leq$ 12 in. (30 cm) while the other stage contains a move of $>$ 12 in. (30 cm), the use of two Controlled Move Sequence Models is required. Each move will be treated as an independent single move. *Example:* Reach to a lever and, without relinquishing control, push it forward 4 in. (10 cm) and then to the side 20 in. (50 cm).

A_1 B_0 G_1 M_1 X_0 I_0 A_0 30

A_0 B_0 G_0 M_3 X_0 I_0 A_0 30

60 TMU

Crank

This category of Move Controlled refers to the manual actions employed to rotate objects such as cranks, handwheels, and reels. These "cranking" actions are performed by moving the fingers, hand, wrist, and/or forearm in a circular path more than half a revolution using one of the patterns pictured in Figure 3.1. Any motion less than half a revolution is not considered a crank and must be treated as a "Push/Pull/Pivot". The overall distance the hand covers when making repetitive circular motions may be larger than any other motions described under the parameter Move Controlled. It is for this reason that a separate column is provided on the Controlled Move data card for Crank.

Figure 3.1 Example of Crank.

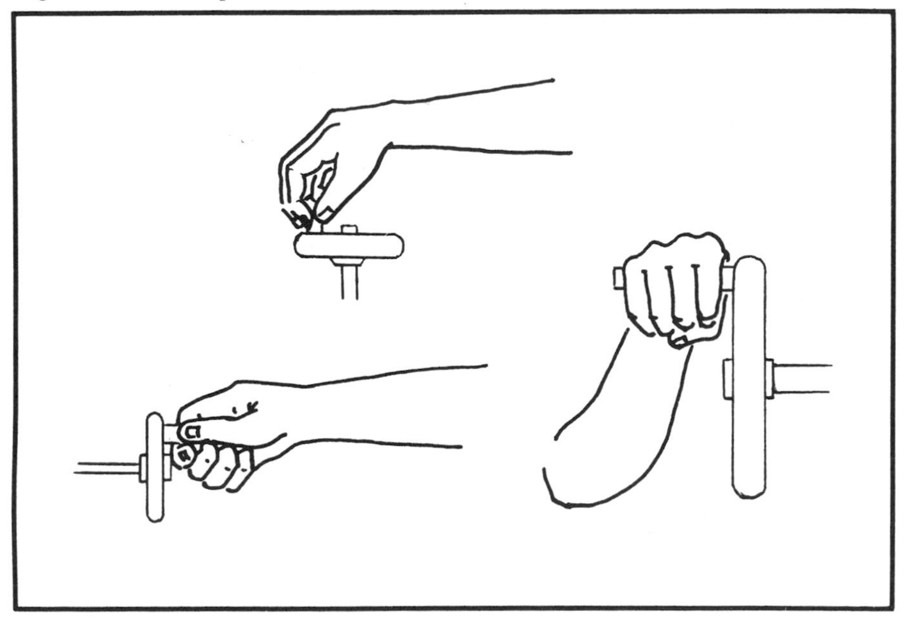

In addition to the actual "cranking time," index values for Crank also include a factor that covers the actions that sometimes occur preliminary to or following the cranking motion. These actions may involve the application of muscular force to seat or unseat the crank or the short manual actions employed to engage or disengage the device undergoing the cranking motion. Table 3.2 lists index values for cranking based on the number of revolutions completed, rounded to the nearest whole number. (Table values are read "up to and including.") *Examples:* Move an engine lathe carriage by cranking a handwheel. Drill a hole in a wooden block by cranking the handle on a manual hand drill.

Process Times (X)

Process time occurs as that portion of work controlled by processes or machines and not by manual actions. The X parameter of the Controlled Move Sequence is intended to cover predominantly fixed process times of relatively short duration. Longer and variable process times such as machining times based on feeds and

Table 3.2

CRANK	
INDEX	REVS.
M_1	-
M_3	1
M_6	3
M_{10}	6
M_{16}	11
M_{24}	16
M_{32}	21
M_{42}	28
M_{54}	36

speeds will normally be considered separately. Table 3.3 lists index values for process times based on the actual clock time (in seconds, minutes, or hours) during which the machine process takes place. The X parameter is indexed by selecting, from the table, the appropriate index value that corresponds to the observed or calculated "actual time." The times listed in Table 3.3 are inclusive. The upper index range limits for each index number (those times appearing on the data card) were calculated in TMU or 1/100,000 hours; therefore, some rounding was needed to determine the upper limit of each index range in terms of the larger, more convenient units (seconds, hours, etc.). For example, the calculated upper limit of index number 6 is 77 TMU, which equals 2.8 seconds; on the data card this time was rounded downward to the nearest half-second (2.5 seconds).

Note: The actual clock time is never placed on the X parameter of the sequence model. Only the index number which statistically represents the actual time should be placed in the sequence model.

Table 3.3 Index Values for Process Time (X)

PROCESS TIME (X)			
Index	Seconds	Minutes	Hours
1	.5	.01	.0001
3	1.5	.02	.0004
6	2.5	.04	.0007
10	4.5	.07	.0012
16	7.0	.11	.0019
24	9.5	.16	.0027
32	13.0	.21	.0036
42	17.0	.28	.0047
54	21.5	.36	.0060
67	26.0	.44	.0073
81	31.5	.52	.0088
96	37.0	.62	.0104
113	43.5	.72	.0121
131	50.5	.84	.0141
152	58.0	.97	.0162
173	66.0	1.10	.0184
196	74.5	1.24	.0207
220	83.5	1.39	.0232
245	92.5	1.54	.0257
270	102.0	1.70	.0284
300	113.0	1.88	.0314
330	124.0	2.06	.0344

Examples of process time: After pushing the button, a process time of 6 seconds occurs while a photocopy machine produces a copy. After throwing a switch, a warm-up period exists in a cathode ray tube. After hitting the palm buttons, a punch press cycles in 1.5 seconds.

Align (I)

Align refers to manual actions following the Move Controlled or at the conclusion of process time to achieve an alignment or specific orientation of objects.

Normally, any adjusting motions required during a controlled move are covered within the M_3 parameter variant for High Control. That index value, however, is not sufficient to cover the activity to line an object up to one or more points following the Controlled Move. This type of alignment is influenced by the ability (or inability) of the eyes to focus on the point(s) simultaneously.

The average area covered by a single eye focus is described by a circle 4 in. (10 cm) in diameter at a normal reading distance of about 16 in. (40 cm) from the eyes (Figure 3.2). Within this "area of normal vision," the alignment of an object to those points can be performed without any additional "eye times." If one of the two points lies outside this area, two separate alignments are required, due to the inability of the eyes to focus on both points simultaneously. In fact, an object would first be aligned to one point, the eyes would then shift to allow the alignment to the second point, and then the object would be finally adjusted to correct for the minor shifting from the first point. The area of normal vision is therefore the basis for the Align Object parameter variants.

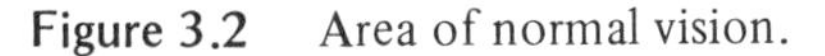

Figure 3.2 Area of normal vision.

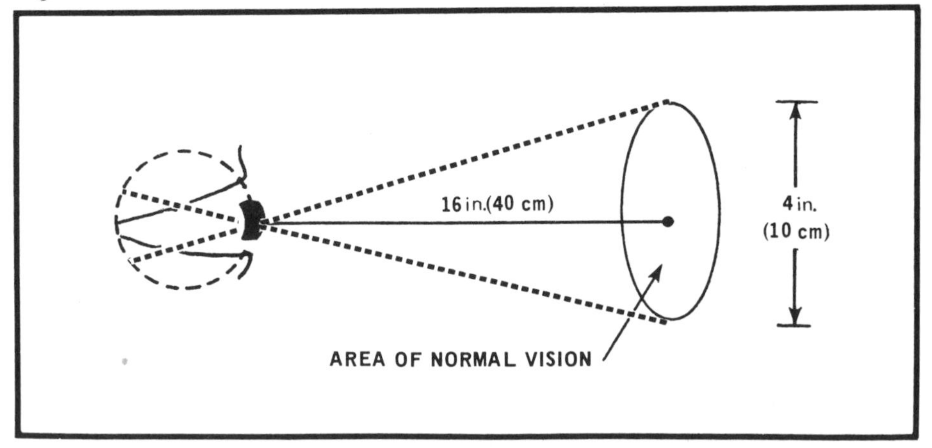

I_1 To One Point

Following a Controlled Move, an object is aligned to one point. Use when demand for a precise alignment is modest and can be satisfied with a single correcting action. This variant is similar to the P_1 variant with the exception that it occurs following a Controlled Move while the P_1 occurs following a General Move. *Example:* Align a mark on a plank to a tablesaw blade.

I_3 To Two Points $\leqslant$ 4 in. (10 cm) Apart

The object is aligned to points not more than 4 in. (10 cm) apart following a Controlled Move. For example, a straightedge is aligned with two marks located 3 in. (7.5 cm) apart, as shown in Figure 3.3 (left). Both points are within the area of normal vision. An increasing demand for precision occurs in this situation. This also includes the time to make more than one correcting motion of the object within the area of normal vision.

I_6 To Two Points $>$4 in. (10 cm) Apart

The object is aligned to points more than 4 in. apart following a Controlled Move. For example, a straightedge is aligned with two marks located 8 in. (20 cm) apart, as shown in Figure 3.3 (right). One point is outside the area of normal vision; therefore, additional eye time must be allowed. Several correcting mo-

Figure 3.3 Align an object to points located $\leqslant$ 4 inches (10 cm) apart (left) and $>$ 4 inches (10 cm) apart (right).

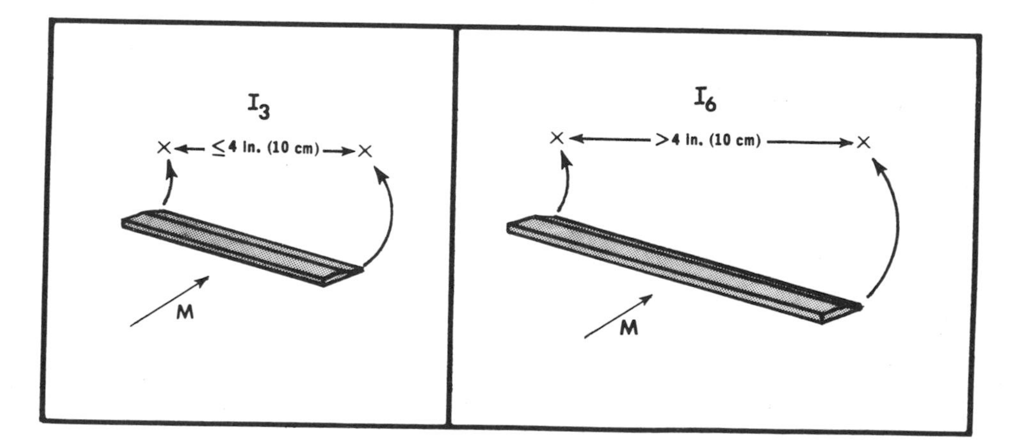

tions and eye focuses are included to allow the time for the hand-eye coordination to be accomplished.

I_{16} Precision

The object is aligned to several points with extreme care or precision following a Controlled Move. For example, the actions to align a french curve or a drawing template with several points will require an I_{16}.

Whenever a Controlled Move involves the Align Object activity, the preceding M parameter is used to describe only the distance the object travels, either $\leqslant$ 12 in. (30 cm) (M_1) or > 12 in. (30 cm) (M_3).

Also, the Align (I) parameter applies only when an alignment occurs following a Controlled Move. Should an object be moved spatially and then be "aligned to two points," the General Move place (P) parameter will account for that time. In fact, a direct relationship between the General Move and the Controlled Move activities should be seen at this time. That relationship is: I:M:: P:A. The alignment (I) of an object occurs after the object is moved over a controlled path (M) and accounts for the time to orient and/or align the object, just as the placement (P) of an object occurs after the spatial displacement of an object (action distance A) and accounts for the time to orient and/or position the object.

Machining Operations

A special group of Align parameter variants is frequently encountered in machine shop operations. Dealing with the alignment of "machining tools," these parameter variants cover the activity following the cranking action (M) to locate the tool on a cutting machine to the correct cutting position. Since these index values are limited in their application to a specific area, they are omitted from the Controlled Move data card. But, because of their importance in this one area, these index values are presented as supplementary data in Table 3.4 and defined below.

I_3 To Workpiece

The machining tool is aligned to the workpiece prior to making a cut. Following any cranking actions (M) to locate the tool near the cutting position, the crank or handwheel is manipulated so that the cutting edge of the tool just touches the workpiece.

I_6 To Scale Mark

The machining tool is aligned to a scale mark prior to making a cut. Following any cranking actions (M) to locate the tool near the cutting position, several taps on the fist of the hand (holding the handwheel) using the other hand are performed to line up the cutting edge of the tool with a scale mark.

Table 3.4 Align Machining Tools

ALIGN	
INDEX	MACHINING TOOL
I_3	TO WORKPIECE
I_6	TO SCALE MARK
I_{10}	TO INDICATOR DIAL

I_{10} To Indicator Dial

The machining tool is aligned to the correct indicator dial setting prior to making a cut. Following any cranking actions (M) to locate the tool near the cutting position, the machine operator must visually locate the indicator dial, read the indicator setting, and carefully adjust the tool to the correct setting by tapping the hand that holds the handwheel several times with the other hand.

Summary of Foot Motions

Movement of the foot could appear in a Controlled Move Sequence Model under the Action Distance (A), the Gain Control (G) or the Move Controlled (M) parameter. A summary follows.

Activity	Analysis
Foot to pedal (without displacing the trunk of the body)	A_1
Take one step	A_3
Gain control of pedal	G_1
Push pedal $\leq$ 12 in. (30 cm)	M_1
Push pedal $>$ 12 in. (30 cm) or with resistance	M_3
Manipulate pedal with High Control (engage clutch)	M_3

Controlled Move Examples

1. From a position in front of a lathe, the operator takes two steps to the side, turns the crank two revolutions, and sets the machining tool against a scale mark.

$A_3 \quad B_0 \quad G_1 \quad M_6 \quad X_0 \quad I_6 \quad A_0$

$(3+1+6+6) \times 10 = 160$ TMU

2. A milling cutter operator walks four steps to the quick-feeding cross lever and engages the feed. The machine time following the 4-in. (10-cm) lever action is 2.5 seconds.

$A_6 \quad B_0 \quad G_1 \quad M_1 \quad X_6 \quad I_0 \quad A_0$

$(6+1+1+6) \times 10 = 140$ TMU

3. A material handler takes hold of a heavy carton with both hands and pushes it 18 in. (45 cm) across conveyor rollers.

$A_1 \quad B_0 \quad G_3 \quad M_3 \quad X_0 \quad I_0 \quad A_0$

$(1+3+3) \times 10 = 70$ TMU

4. Using the foot pedal to activate the machine, a sewing machine operator makes a stitch requiring 3.5 seconds process time. (The operator must reach the pedal with the foot.)

$A_1 \quad B_0 \quad G_1 \quad M_1 \quad X_{10} \quad I_0 \quad A_0$

$(1+1+1+10) \times 10 = 130$ TMU

4
The Tool Use Sequence

Multiple Moves

Occasionally, an activity will contain a pattern of several Controlled Moves or combination of General and Controlled Moves in succession. These multiple moves, or actions, are frequently encountered when fastening or loosening threaded fasteners using either the hand or hand tools such as screwdrivers, wrenches, or ratchets. Special parameter variants and a specially developed Tool Use Sequence describe these multiple moves under the parameters of Fasten/Loosen in terms of the body member performing the action, i.e., finger, wrist, or arm. For example, running a nut down with the fingers would be considered a finger action, while tightening a wood screw with a screwdriver would require a wrist action. These actions are by literal definition a series of Controlled Moves. Remember that any activity can be broken down into a series of General and/or Controlled Moves, even the use of common hand tools, e.g., get and position screwdriver (General Move), or tighten screw (a series of Controlled Moves). However, as stated earlier, the Tool Use Sequence Model was developed to simplify the analysis of activities related to the use of common hand tools. Because of the ease of use, the consistency provided, and the analysis time saved, such sets of multiple moves can normally be analyzed using the Tool Use Sequence, with the activity being described in terms of fastening, loosening, cutting, etc. Therefore, situations that appear to be an extended series of General and/or Multiple Controlled Moves, such as using a wrench, pencil, screwdriver, or micrometer, are better analyzed using the Tool Use Sequence Model.

The development of the Tool Use Sequence Model not only increased consistency and application speed, but it also provided analyses that were more accurate than those using a series of sequence models to analyze the use of tools. It was discovered that repeating analyses compounded deviations that existed between the allowed time (assigned index number) and the "actual time," whereas developing values using the statistically determined index ranges and assigning one index number, representing Tool Use, eliminated the compounding of any deviation. Accuracy was therefore maintained through the system design and was independent of the nature or complexity of the manual actions performed (see Appendix A, Theory). Wherever possible, develop (see Chapter 7) and/or use existing Tool Use data when analyzing the use of common hand tools.

Manual work is not always performed with the hand alone. The use of tools extends the strength and capabilities of the hand through leverage. Even though much mechanization has occurred in industry, a large and very critical portion of work still remains literally in the hands of the worker. Because of the desirability of having the MOST Work Measurement Technique apply to all manual work and since analysis of the frequent use of certain tools through a series of General and Controlled Moves could be time consuming and result in inconsistent applications, a third manual sequence model was developed—the Tool Use Sequence Model.

The Tool Use Sequence is composed of subactivities from the General Move Sequence, along with specially designed parameters describing the actions performed with hand tools or, in some cases, the use of certain mental processes. Tool Use follows a fixed sequence of subactivities occurring in five main activity phases:

1. Get object or tool.
2. Place object or tool in working position.
3. Use tool.
4. Put aside object or tool.
5. Return to workplace.

The Sequence Model

These five activity phases form the basis for the activity sequence describing the *handling and use of hand tools*. The sequence model takes the form of a series of letters representing each of the various subactivities of the Tool Use Activity Sequence:

GET OBJECT OR TOOL	PLACE OBJECT OR TOOL	USE TOOL	ASIDE OBJECT OR TOOL	RETURN
A B G	A B P		A B P	A

where A = Action distance
B = Body motion
G = Gain control
P = Place

The space in the sequence model (use tool) is provided for the insertion of one of the following Tool Use parameters. These parameters refer to the final results of using the tool and are:

F = Fasten
L = Loosen
C = Cut
S = Surface treat
M = Measure
R = Record
T = Think

Parameter Definitions

F Fasten

This parameter refers to mechanically assembling one object to another, using the fingers, a hand, or a hand tool.

L Loosen

This parameter refers to mechanically disassembling one object from another using the fingers, a hand, or a hand tool.

C Cut

This parameter describes the manual actions employed to separate, divide, or remove part of an object using a sharp-edged hand tool.

S Surface Treat

This parameter covers the activities aimed at removing unwanted material or particles from, or applying a substance, coating, or finish to, the surface of an object.

M Measure

This parameter refers to the actions employed in determining a certain physical characteristic of an object by comparison with a standard measuring device.

R Record

This parameter covers the manual actions performed with a pencil, pen, chalk, or other marking tool for the purpose of recording information.

T Think

This parameter refers to the eye actions and mental activity employed to obtain information (read) or to inspect an object.

Parameter Indexing

With the exception of the special use tool parameters, the Tool Use Sequence Model contains only parameters from the General Move Sequence. Index values for these parameters are of course found in the table of General Move data (Table 2.1). Two additional tables are provided for the use tool parameters. Table 4.1 contains index values for tools dealing with the parameters fasten or loosen, while Table 4.2 covers activities such as cutting, cleaning, gauging, reading, and writing. The use of these two tables for indexing the use tool parameter follows the same procedure outlined previously in the General and Controlled Move chapters.

Consider, for example, an assembly operation in which a bolt is used to fasten one object to another. First, the operator picks up a bolt from a bin located within reach and places it in the required location. It is then run down with three spins with the fingers. The sequence model would be indexed:

$A_1 \quad B_0 \quad G_1 \quad A_1 \quad B_0 \quad P_3 \quad F_6 \quad A_0 \quad B_0 \quad P_0 \quad A_0$

$(1+1+1+3+6) \times 10 = 120$ TMU

In this example, the "get tool" and "place tool" sections of the sequence model are used for getting and placing the bolt; placement of a threaded fastener will nearly always be a P_3 (with adjustments) unless it takes place in a blind or obstructed location (P_6). Since this is a fastening activity, the F parameter is chosen and inserted in the sequence model. The appropriate index value is determined by considering the body member predominantly performing the fastening activity (in this case, the fingers) and the number of actions performed. Looking at the Fasten/Loosen table (Table 4.1), we see that up to three finger actions requires an index value of 6. The remaining parameters in the sequence (A, B, P, and A) carry zero index values, since no activity was performed to set aside a tool or object.

In the second part of this example, let us say that following the above fastening activity, the operator then picks up a small box-end wrench (lying on the workbench within reach) and tightens the bolt with three wrist actions. This second sequence model would be indexed:

$A_1 \quad B_0 \quad G_1 \quad A_1 \quad B_0 \quad P_3 \quad F_{10} \quad A_1 \quad B_0 \quad P_1 \quad A_0$

$(1+1+1+3+10+1+1) \times 10 = 180$ TMU

Again using the Fasten/Loosen table, the index value in this case is taken from the reposition column below wrist actions. Index values in this column refer to

Table 4.1 Tool Use–Fasten/Loosen*

INDEX	ABGABP ABPA					TOOL USE SEQUENCE						INDEX
	Fasten or Loosen											
	FINGER ACTION	WRIST ACTION				ARM ACTION				TOOL ACTION		
	SPINS	TURNS	STROKES (REPOSITION)	CRANKS	TAPS	TURNS	STROKES (REPOSITION)	CRANKS	STRIKES	SCREW DIAMETER		
	FINGERS SCREW-DRIVER	HAND SCREW-DRIVER RATCHET T-WRENCH	WRENCH ALLEN KEY	WRENCH ALLEN KEY RATCHET	HAND HAMMER	RATCHET	WRENCH ALLEN KEY	WRENCH ALLEN KEY RATCHET	HAND HAMMER	POWER WRENCH		
1	1	-	-	-	1	-	-	-	-	-		1
3	2	1	1	1	3	1	1	-	1	1/4'' 6 mm		3
6	3	3	2	3	6	2	-	1	3	1'' 25 mm		6
10	8	5	3	5	10	4	2	2	5			10
16	16	9	5	8	16	6	3	3	8			16
24	25	13	8	11	23	9	4	5	12			24
32	35	17	10	15	30	12	6	6	16			32
42	47	23	13	20	39	15	8	8	21			42
54	61	29	17	25	50	20	10	11	27			54

* Values are read "up to and including."

the way in which a wrench is normally used. That is, following each wrist action, the wrench must be "repositioned" on the fastener before making any subsequent actions. In our example three wrist actions are performed with the wrench. The corresponding index value is therefore F_{10}.

In addition to the use tool section of the sequence model, the remaining parameters in this sequence apply to the handling of the tool. The P_3 prior to use tool in the example covers the initial placement of the wrench on the bolt. The parameters following the use tool section, i.e., A_1 B_0 P_1 A_0, indicate that the wrench is placed aside following the fastening activity.

Use of the other use tool data (Table 4.2) can be demonstrated with a third example. Suppose that during a sewing operation a seamstress picks up a pair of scissors and makes three cuts to remove the excess material from around a stitch. This activity would be described as follows.

A_1 B_0 G_1 A_1 B_0 P_1 C_6 A_1 B_0 P_1 A_0

$(1+1+1+1+6+1+1) \times 10 = 120$ TMU

The appropriate use tool parameter for this example would be Cut which is represented by the letter C. Looking down the column headed Cut in Table 4.2, we see that "three cuts with scissors" carries the index value C_6. The initial placement of the scissors prior to the cutting action is assumed to be P_1 in this case.

The remainder of this chapter will examine in detail each of the use tool parameters and discuss their application.

The Fasten/Loosen Data Card

Fasten or loosen refers to mechanically assembling or disassembling one object to or from another using the fingers, a hand, or a hand tool. Index values for the F and L parameters are primarily grouped according to the body member (e.g., finger, wrist, or arm) predominantly performing the tool action. An additional category is provided for power-operated hand tools.

With the exception of power tools, all the data found in Table 4.1 refers to the number of actions performed by the respective body member during either a fastening (F) or loosening (L) activity. An action is defined as the back-and-forth movement of the fingers, wrist, or arm to perform one "stroke," "pull" or "turn" with the tool. In the case of the crank data, action refers to one revolution of the tool.

Finger Actions (Spins)

Finger action refers to the movements of the fingers and thumb to run a threaded fastener down or out. These short finger movements are characterized by rolling or spinning an object between the thumb and index finger. Examples would include running a nut down with the fingers or turning a machine screw with a small screwdriver. Because of the limited strength in the fingers, the mus-

Table 4.2 Tool Use Data Card*

ABGABP ABPA — TOOL USE SEQUENCE

INDEX	C				S			M	R			T			INDEX
	GRIP, ETC.	CUTOFF	CUT	SLICE	AIR-CLEAN	BRUSH-CLEAN	WIPE	MEASURE	WRITE		MARK	INSPECT	READ		
	PLIERS		SCISSORS	KNIFE	NOZZLE	BRUSH	CLOTH	MEASURING DEVICE	PENCIL		MARKER	EYES FINGERS	EYES		
		(WIRE)	CUTS	STROKES	SQ. FT. 0,1 M²	SQ. FT. 0,1 M²	SQ. FT. 0,1 M²	IN. (CM) FT. (M)	DIGITS	WORDS	DIGITS	POINTS	DIGITS SINGLE WORDS	TEXT OF WORDS	
1	GRIP		1	-	-	-	-		1	-	CHECK MARK	1	1	3	1
3		SOFT	2	1	-	-	1/2		2	-	1 SCRIBE LINE	3	3 GAUGE	8	3
6	TWIST BEND LOOP	MEDIUM	4		1 POINT OR CAVITY	1 SMALL OBJECT	-		4	1	2	5 TOUCH FOR HEAT	6 SCALE VALUE DATE/TIME	15	6
10		HARD	7	3	-	-	1	PROFILE-GAUGE	6	-	3	9 FEEL FOR DEFECT	12 VERNIER-SCALE	24	10
16			11	4	3	2	2	FIXED-SCALE CALIPER 12 IN. (30 CM)	9 SIGNATURE DATE	2	5		TABLE VALUE	38	16
24			15	6	4	3	-	FEELER-GAUGE	13	3	7			54	24
32			20	9	7	5	5	STEEL-TAPE 6FT. (2M) DEPTH MICROMETER	18	4	10			72	32
42			27	11	10	7	7	OD-MICROMETER 4 IN. (10 CM)	23	5	13			94	42
54			33					ID-MICROMETER 4 IN. (10 CM)	29	7	16			119	54

*Values are read "up to and including."

cular force pressure exerted on the fastener while performing spins is minimal. The finger action data in Table 4.1, however, do include a light application of pressure seating and unseating of the fastener, as in replacing a cap on a bottle.

Wrist Actions

A wrist action refers to the twisting motion of the wrist about the axis of the forearm, or the pivoting of the hand from the wrist with either a circular or back-and-forth motion. Wrist actions normally include hand movements up to 6 in. (15 cm) in length as measured from the knuckle at the base of the index finger. As Table 4.1 indicates, data are classified according to the manner in which the wrist actions are performed.

Wrist Turn Tools covered under the heading wrist turns include using the hand, screwdriver, ratchet, or small T-wrench. While used, these tools are not removed from the fastener and are not repositioned on the fastener following each action. The time for wrist turns *does* include the time for repositioning the hand or tool handle after each action. Also, as a result of the added strength possible when using the larger muscles of the hand and forearm, a final tighten or initial loosen can be accomplished with a wrist turn. Therefore, the index values assigned from the wrist turn column also include the time for final tightening or initial loosening of a fastener.

Wrist Reposition (Stroke) The wrist reposition column refers to the method normally employed when using a wrench. That is, following each stroke with the tool, the wrench must be removed from and repositioned on the fastener before making each subsequent stroke. Index values in this column apply to the number of power strokes (actions) performed with the wrench (*not the number of repositions of the wrench*). Included also is the time for the wrench to be removed from and repositioned on the fastener between the strokes. Like wrist turn, the data for wrist reposition also allow for the final tightening or initial loosening activity. Tools covered by this parameter include fixed-end, adjustable, and Allen wrenches, etc. (tools that are normally "repositioned" on a fastener for use).

Wrist Crank Data from the wrist crank column apply to tools that are spun or rotated around a fastener while remaining affixed to it. They are guided with a circular movement of the hand as it is pivoted from the wrist (Fig. 4.1). This type of wrist action is sometimes used with either wrenches or ratchets when there are no obstructions in the path of the tool. Following the initial placement of the tool, the fingers and hand are used to push or "crank" the tool completely around the fastener. However, these wrist actions are employed by operators only when little or no resistance is encountered; therefore, data in the wrist crank column *do not* include the time for final tightening or initial loosening of a fastener. If, following a number of wrist cranks, a fastener is final tight-

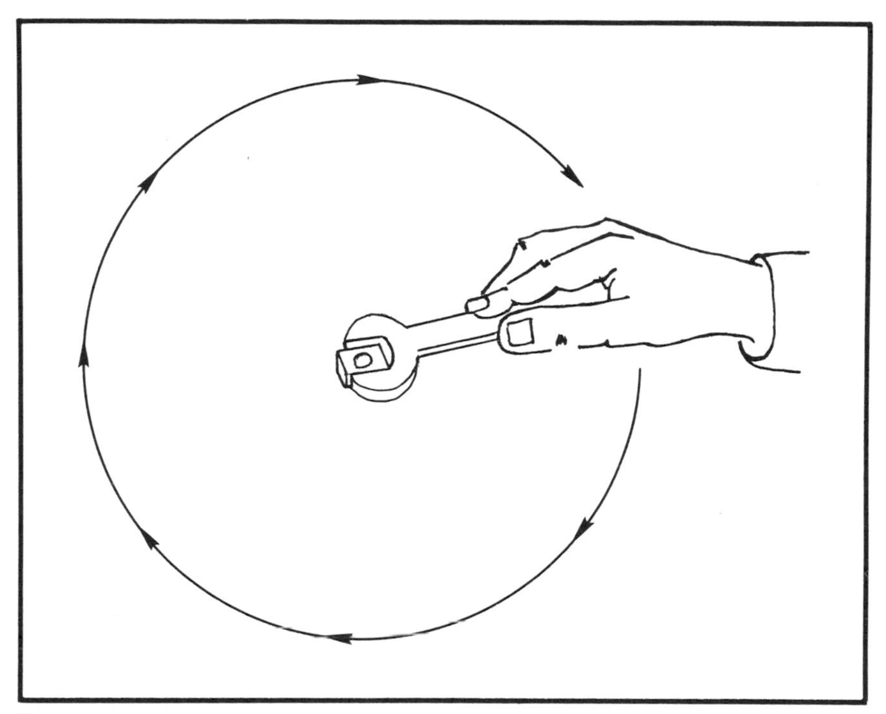

Figure 4.1 Wrist crank.

ened, the normal type of tool action (wrist turn or wrist reposition) will be used to analyze the final tightening activity. Usually, one or several of these actions will be needed. Index values for wrist crank refer to the number of revolutions performed with the tool.

Fasten/Loosen with continuous cranking motions is the most economical way of running down a screw, due to the fact that one cranking motion results in running down one thread on the screw while other methods produce one-third or one-sixth thread per action.

Tap The use of a tool such as a small hammer is covered by the data under the heading tap. Index values from the tap column refer to the short tapping motions performed with the hand as it is pivoted at the wrist. Data in this column refer to the number of tapping actions made with the hand.

Arm Actions

Arm actions refer to the motions of the hand requiring elbow and shoulder movements. With the wrist relatively rigid, the forearm is pivoted from the elbow with either an up-and-down, circular, or back-and-forth of motion. These forearm motions may be assisted by the pivoting of the upper arm from the shoulder. Arm actions cover hand movements of between 6 and 18 in. (15 and 45 cm) in length or a circular motion with a diameter up to 24 in. (60 cm).

Arm Turn In the first column, the tools covered under the heading arm turns include only the use of a ratchet. Arm actions of this type are employed when the ratchet is held near the end of the handle, resulting in a pulling action on the tool. Index values from the arm turn column include time for the final tightening or initial loosening that may occur in the complete fastening or loosening activity.

Arm Reposition (Stroke) Similar to the wrist reposition data, the arm reposition column applies to the normal method of using a wrench. That is, following each stroke or pull with the tool, the wrench must be removed and repositioned again on the fastener before making a subsequent pull. Index values in this column apply to the number of arm actions (pulls) performed with the wrench (not the number of repositions). Index values for arm reposition allow for the final tightening or initial loosening activity that may occur in the complete fastening or loosening activity. Tools covered by this parameter include fixed-end, adjustable, and Allen wrenches.

Arm Crank The data from the arm crank column apply to tools used with a circular movement of the forearm as it is pivoted at the elbow or the shoulder (Fig. 4.1). Arm actions of this type are occasionally used with either wrenches or ratchets when there are no obstructions in the path of the tool. The hand is used to push or crank the tool around the fastener. Like the wrist actions under the same heading, this type of action is employed only when resistance is minimal: therefore, the values in the arm crank column *do not* include the time for final tightening or initial loosening of a fastener. The data in this column refer to the number of revolutions performed with the tool; if a partial revolution is observed, round to the nearest whole number.

Strike The use of a hammer with arm actions is accounted for under the heading strike. The data in this column refer to the up-and-down motions performed with the hand as it is pivoted from the elbow.

T-wrench, Two Hands The following supplementary data (not found on the Fasten/Loosen data card) are provided to analyze the use of a large T-wrench with two hands (see Table 4.3). Each arm action involves a 180-degree turn of the T-wrench. All subsequent two-handed arm actions required that each hand reach to the opposite handle before making the next turn. The data for two hands allow for the final tightening or initial loosening involved in the complete fastening or loosening activity. This would also be appropriate for turning a valve or other such item with two hands.

Table 4.3 Supplementary Data, for Two Hands

2 HANDS	
INDEX	NUMBER OF ARM ACTIONS
1	-
3	-
6	1
10	-
16	3
24	6
32	8
42	11
54	15

Power Tools

Power tools refer to the use of power-operated hand tools. The data provided in Table 4.1 cover electric and pneumatic power wrenches. Index values are based on the time required to run a standard threaded fastener down or out, a length equal to 1 or 2 times the screw diameter of the fastener (Fig. 4.2), the distance to hold a nut securely. Two values are found in the table: F_3 or L_3 for a screw diameter of 1/4 in. (6 mm) or smaller, and F_6 or L_6 for larger screws up to and including 1 in. (25 mm) in diameter. Therefore, to apply F or L to a power tool, simply choose the fasten or loosen value based on the diameter of the fastener. *Note:* This applies to standard fasteners (where the length of holding threads is 1-2 times the diameter) only.

When running down or out longer fasteners, where more threads are needed to hold the item or threads are fine, a frequency can be applied to the F or L value chosen. For example, run in a 3/8 in. diameter bolt 1 in. Normally, the bolt would be run in approximately 3/4 in. or 2 times its diameter, and an F_6 would be appropriate. However, in this case it is being run in 3 to 4 times its diameter. An F_6 times two (F_6) (2) is now appropriate.

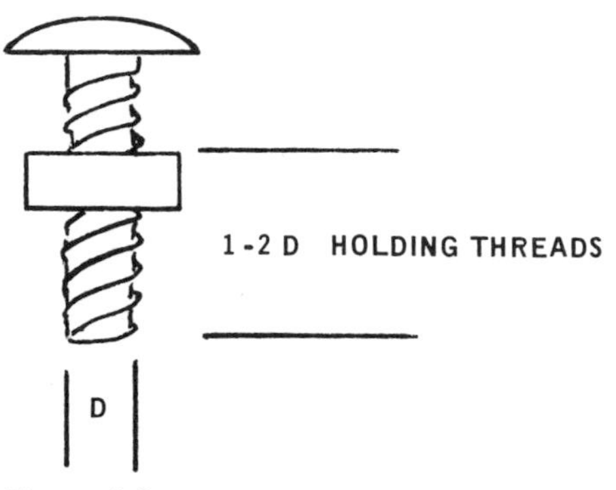

Figure 4.2

Also, it must be remembered that the basic values for Fasten/Loosen with a power tool must be compared to the brands of power tools used in the plant. Should there be a difference in the basic MOST values for Fasten/Loosen with a power tool and those studied, new index values for the tools on hand must be created using the Index Value Determination Form (the procedure for developing index values is outlined in Chap. 7).

Supplementary values for special tools or special situations not found on the data card (Table 4.1) have been developed and are included below.

F_6 Torque Wrench

Tighten a bolt or nut with a torque wrench having a handle length of up to 10 in. (25 cm). The value is for one arm action and includes the time either to align the dial or to await the click.

F_{10} Torque Wrench

Tighten a bolt or nut with a torque wrench having a handle length of 10 to 15 in. (25-38 cm). The value is for one arm action and includes the time either to align the dial or to await the click.

F_{16} Torque Wrench

Tighten a bolt or nut with a torque wrench having a handle length of 15 to 40 in. (38 cm to 1 m). The value is for one arm action and includes the time either to align the dial or to await the click.

Tool Placement

The P parameter prior to use tool covers the setting of the tool or object in the working position prior to the tool action. The index value for P (placement of the tool) should be selected using the guidelines set forth in the General Move chapter. However, as a general rule, the P parameter for the Fasten/Loosen tools will carry the index values indicated in Table 4.4. Notice that placing the tool, finger, or hands on a fastener is considered a P_1. This is, of course, a G_1 Gain Control in actuality. However, since the fingers or hands are used in the same way as a fastening or loosening tool, the activity is considered the placement of a tool instead of a grasp. For example,

GET TOOL	PLACE TOOL	USE TOOL	ASIDE TOOL	RETURN	
$A_0 B_0 G_0$	$A_1 B_0 P_1$	F_6	$A_0 B_0 P_0$	A_0	80 TMU

If the fingers or hands are concerned with placing a fastener, such as a nut or bolt, immediately preceding the action to fasten it, the P parameter refers to the placement of the fastener. The placement of a threaded fastener nearly always requires a P_3 placement unless the placement occurs in a blind or obstructed location; under those conditions, a P_6 would be appropriate, for example,

GET FASTENER	PLACE FASTENER	USE TOOL	ASIDE TOOL	RETURN	
$A_1 B_0 G_1$	$A_1 B_0 P_3$	F_6	$A_0 B_0 P_0$	A_0	120 TMU

Notice also from Table 4.4 that the placement of an adjustable wrench occurs with a P_6. This larger index value is required to cover the additional actions necessary to adjust the jaws of the wrench (with intermediate moves) to the size of the fastener. A value of P_3 would be sufficient if the wrench had been previously adjusted to the proper fastener size.

There may or may not be an initial placement of a hammer prior to any tapping or striking actions. Normally, if a hammer is being used to drive small nails or tacks, the hammerhead will be positioned over the nail (P_1) prior to performing any actions. In many cases, however, no initial placement of the hand or hammer is necessary (P_0), for example, before simply tapping or striking a larger object or surface area.

Tool Use Frequencies

Occasionally an activity may involve the fastening or loosening of several fasteners in succession using the same tool. By using a "special convention," the entire activity can normally be described using only one Tool Use Sequence Model. For example, an operator picks up a screwdriver within reach and tightens two screws with six wrist turns each and then sets aside the screwdriver. The first step in making an analysis of this activity is to analyze the situation as if only one screw were fastened and then repeat the appropriate parameters to tighten the second screw for example,

$A_1 \quad B_0 \quad G_1 \quad A_1 \quad B_0 \quad P_3 \quad F_{16} \quad A_1 \quad B_0 \quad P_1 \quad A_0$ (For one screw)

.4 Index Values for Tool Placement

TOOL	INDEX VALUE
HAMMER	P_0 (P_1)
FINGERS OR HAND	P_1 (P_3 OR P_6)
KNIFE	P_1 (P_3)
SCISSORS	P_1 (P_3)
PLIERS	P_1 (P_3)
WRITING INSTRUMENT	P_1
MEASURING DEVICE	P_1
SURFACE TREATING DEVICE	P_1
SCREWDRIVER	P_3
RATCHET	P_3
T-WRENCH	P_3
FIXED END WRENCH	P_3
ALLEN WRENCH	P_3
POWER WRENCH	P_3
ADJUSTABLE WRENCH	P_6

What must be repeated to fasten the second screw? First, there is a reach over to the second screw, then the tool must be positioned, and then fastened; therefore, the placement and the fastening or loosening must be repeated.

To cover the action distance of the tool to each fastener requires that an A parameter be written into the sequence model between the P and either the F or L parameters. For example,

Add an "A" to cover the Reach between the fasteners

$$A_1 \quad B_0 \quad G_1 \quad A_1 \quad B_0 \quad P_3 \quad A \quad F_{16} \quad A_1 \quad B_0 \quad P_1 \quad A_0$$

Parentheses are then placed around all those parameters to be repeated (e.g., then P, A, and F or L). For example,

Add parentheses

$$A_1 \quad B_0 \quad G_1 \quad A_1 \quad B_0 \quad (P_3 \quad A \quad F_{16}) \quad A_1 \quad B_0 \quad P_1 \quad A_0$$

If the distance between the screws is $\leqslant 2$ in. (5 cm), an A_0 is placed between the P and F or L parameters. For example, using a screwdriver, tighten two screws with six wrist turns each. The distance between the screws is $\leqslant 2$ in. (5 cm).

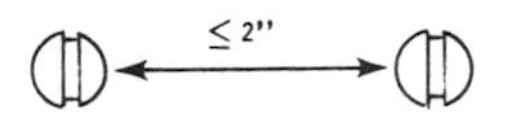

$A_1 \; B_0 \; G_1 \; A_1 \; B_0 \; (P_3 \; A_0 \; F_{16}) \; A_1 \; B_0 \; P_1 \; A_0$ (2) = 430 TMU

Note: "A" must be added to the use tool section to account for the distance between the screws.

If the distance between the screws is >2 in. (5 cm), an A_1 must be placed in the parentheses. Since the action distance to each fastener is covered by the A parameter within the parentheses, *the A following Gain Control will now carry a zero index value.* This is to avoid counting an "extra" action distance value. For example, using a screwdriver, tighten two screws with six wrist turns each. The distance between the screws is 5 in. (12.5 cm).

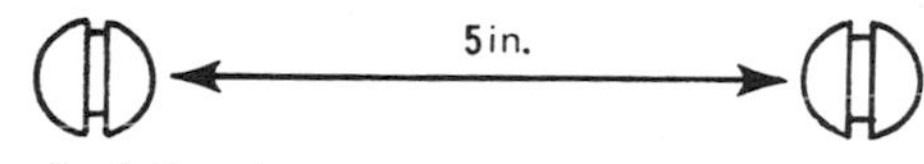

The incorrect time calculation is:

$A_1 \; B_0 \; G_1 \; A_1 \; B_0 \; (P_3 \; A_1 \; F_{16}) \; A_1 \; B_0 \; P_1 \; A_0$ (2) = 450 TMU

Note: When the distance between fasteners is >2 in. (5 cm) you must drop the placement A_1 value since it will be included in the frequency value. As illustrated below, there are two action distances, one to the first screw and one to the second. The number in the frequency column times the A in the parentheses will account for all the reaches needed.

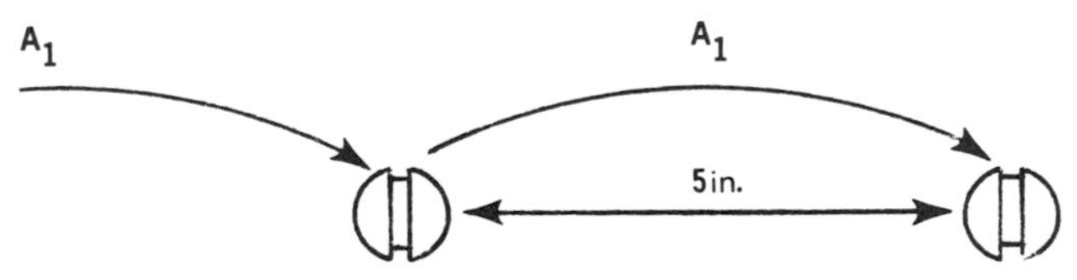

The multiplier for these parameters (the number of fasteners included in the fastening or loosening activity) is then placed in the frequency column of the MOST calculation sheet, also within parentheses. For example, the correct time calculation is:

$A_1 \; B_0 \; G_1 \; A_0 \; B_0 \; (P_3 \; A_1 \; F_{16}) \; A_1 \; B_0 \; P_1 \; A_0$ (2)

The time calculation for the fastening or loosening activity is performed by simply adding all index values contained within the parentheses and multiplying this sum by the number of fasteners involved (the frequency). The sequence model total is obtained by adding to this the index values from the remaining parameters. The conversion to TMU is obtained in the usual way by multiplying the total by 10. For example,

$A_1 \ B_0 \ G_1 \ A_0 \ B_0 \ (P_3 \ A_1 \ F_{16}) \ A_1 \ B_0 \ P_1 \ A_0 \ (2) = 440$ TMU

$(3+1+16) = 20 \times 2 = 40+1+1+1+1 = 44 \times 10 = 440$ TMU

Multiple Tool Actions

The data found in Table 4.1 are classified according to the body member predominantly performing the tool action, not by the tool itself, the reason being that the tool can be used with more than one type of tool action. In fact, an operator may employ a combination of different finger, wrist, or arm actions during a fastening or loosening activity with a single tool. This may be found quite often when finger actions (spins) or wrist or arm cranks are involved, for the values in those columns on the Fasten/Loosen data card *do not* include the time for final tightening or initial loosening of a fastener. Therefore, as previously explained, when a fastener is final tightened or initial loosened in conjunction with any of the above activities, another activity (e.g., wrist or arm turn) will be performed and should be analyzed.

For example, when using a screwdriver, the initial tool actions to run down a screw may be performed with finger spins if no resistance is encountered. But the final tightening (more than the finger pressure to seat the screw) may require the use of wrist actions. As another example, a ratchet may first be used with cranking actions followed by wrist turns to final tighten the fastener.

These and other similar fastening or loosening activities can be described in one sequence model by placing the appropriate index values for each of the tool actions on a single F or L parameter. Index values for these multiple tool actions are separated by a plus (+) sign. Consider the above example with the screwdriver in which 18 finger actions and four wrist turns are employed to fasten a machine screw.

$A_1 \ B_0 \ G_1 \ A_1 \ B_0 \ P_3 \ F_{24+10} \ A_1 \ B_0 \ P_1 \ A_0$

$(1+1+1+3+24+10+1+1) \times 10 = 420$ TMU

Using the ratchet example, let us say that a nut is run down three revolutions with a wrist crank followed by six wrist turns. The sequence model is indexed

$A_1 \ B_0 \ G_1 \ A_1 \ B_0 \ P_3 \ F_{6+16} \ A_1 \ B_0 \ P_1 \ A_0$

$(1+1+1+3+6+16+1+1) \times 10 = 300$ TMU

This procedure should be used only when two different types of action are performed with the same tool.

Tool Use Examples for Fasten/Loosen

1. Obtain a nut from a parts bin located within reach, place it on a bolt, and run it down with seven finger actions.

A_1 B_0 G_1 A_1 B_0 P_3 F_{10} A_0 B_0 P_0 A_0

(1+1+1+3+10) X 10 = 160 TMU

2. Pick up a small screwdriver from within reach and fasten a screw with five finger actions, final tighten the screw with one wrist turn, and set aside the tool.

A_1 B_0 G_1 A_1 B_0 P_3 F_{10+3} A_1 B_0 P_1 A_0

(1+1+1+3+10+3+1+1) X 10 = 210 TMU

3. Obtain a power wrench from within reach, run down four 3/8 in. (10-mm) bolts located 6 in. (15 cm) apart, and set aside wrench.

A_1 B_0 G_1 A_0 B_0 $\left(P_3 \quad A_1 \quad F_6\right)$ A_1 B_0 P_1 A_0 (4)

[(3+1+6) X (4)+1+1+1+1] X 10 = 440 TMU

4. From a position in front of an engine lathe, obtain a large T-wrench located five steps away and loosen one bolt on a chuck on the engine lathe with both hands using five arm actions. Set aside the T-wrench from the machine (but within reach).

A_{10} B_0 G_1 A_{10} B_0 P_3 L_{24} A_1 B_0 P_1 A_0

(10+1+10+3+24+1+1) X 10 = 500 TMU

5. Obtain a 1/4-in. (6-mm) ratchet located within reach and run a bolt down by rotating the ratchet with eight wrist cranks. Final tighten the bolt with four wrist turns and set aside the ratchet.

A_1 B_0 G_1 A_1 B_0 P_3 F_{16+10} A_1 B_0 P_1 A_0

(1+1+1+3+16+10+1+1) X 10 = 340 TMU

6. Walk five steps to a tool cabinet and get a 12-in. (30-cm) fixed-end wrench. Return to the workplace and loosen two bolts located 12 in. (30 cm) apart using four arm actions. Set the wrench aside, within reach.

A_{10} B_0 G_1 A_{10} B_0 $\left(P_3 \quad A_1 \quad L_{24}\right)$ A_1 B_0 P_1 A_0 (2)

[(3+1+24) X (2)+10+1+10+1+1] X 10 = 790 TMU

The Data Card for Cut, Surface Treat, Measure, Record, and Think

On this data card are found the index values for common activities found within the parameters of cut, surface treat, measure, record, and think. The list of values given are not meant to be comprehensive. In fact, should special or supplementary activities (tools or operations) be required to analyze a particular situation, the analyst is encouraged to develop those values under the guidelines set forth in Chapter 7. With this, the analyst tailors the data card to his or her particular situation or industry.

Cut

Cut describes the manual actions employed to separate, divide, or remove part of an object using a sharp-edged hand tool. As Table 4.2 indicates, index values for the C parameter cover the use of pliers, scissors, or knife for general cutting activities. Each of these cutting tools and their use are described as follows.

Pliers

Three different methods may be employed to cut through wire using pliers. The particular method employed largely depends on the hardness of the wire material and the diameter or gauge of the wire. Small-gauge copper wire, for instance, requires only a squeezing of the hand to simply snip off the wire (soft wire). However, with larger-gauge wire or harder material such as steel, two separate cuts may be required to completely sever the wire (medium wire). That is, following an initial cut, the pliers are rotated around the wire and repositioned over the cut before completely cutting through the wire. A third method may be encountered with the largest gauge and hardest wire (hard wire). In addition to requiring two cuts, both hands are needed to apply sufficient force to cut through the wire.

C_3 **Soft**

This parameter would apply to cutting a soft steel, copper, or other small-gauge wire. Recognized by using the pliers with one hand and making one cut.

C_6 **Medium**

This parameter would apply to cutting a steel wire or cable and can be recognized using the pliers with one hand and making two cuts.

C_{10} **Hard**

This parameter would apply to cutting a heavier wire (approximately 10 gauge) and can be recognized by using two hands and making two cuts.

The data (Table 4.2) for using pliers therefore include three index values for cutting wire. The value used most often in small electrical assembly work would be C_3 (cutoff, soft wire). In heavier assembly work or electrical maintenance, for instance, the appropriate value may be C_6 (cutoff, medium wire) or even C_{10} (cutoff, hard wire). Placement of the pliers is normally a P_3.

Also included under the column for pliers are three common activities performed with the pliers.

C_1 Grip

Following the initial placement of the pliers, the operator squeezes the pliers to simply hold an item and subsequently releases the pressure on the item. *Example:* using pliers, hold a wire in place for soldering.

C_6 Twist

Following the placement of the pliers on two wires, the jaws are closed and two twisting motions of the pliers join the wires together. Should more than two twisting actions need to be analyzed, break the number observed into groups of two and apply a frequency to the C_6. *Example:* using pliers, twist the ends of two wires together.

C_6 Bend Loop

Following the initial placement of the pliers the operator closes the jaws and using two actions bends a loop or eye in the end of a wire. *Example:* using pliers, form an eye in the end of a wire to fit over a terminal in a junction box.

C_{16} Bend Cotter Pin

Following the initial placement, an operator bends both legs on a cotter pin to hold it in position. *Example:* using pliers, bend legs on a cotter pin to secure it through a small shaft. This is a supplementary value developed for using pliers and does not appear on the data card.

Scissors

These data apply to cutting paper, fabric, light cardboard, or other similar material using scissors. Index values are selected according to the number of cuts or scissors actions employed during the cutting activity. To cut off a piece of thread, for example, only one cutting action would be required. Accordingly, the appropriate index value from Table 4.2 would be C_1 (one cut with scissors). Likewise, the actions of a seamstress in cutting through a piece of fabric with four cutting actions would be indexed C_6 (four cuts with scissors). Placement of scissors is normally a P_1 (P_3 if exact placement is required).

Knife

The use of a sharp knife for cutting string or light cord or to cut through corrugated material or cardboard carries the index value C_3. This value also applies to the activities aimed at slicing open a corrugated box. The length of the cut can be $>$ 18 in. (45 cm). If the box is wrapped with string or cord, the initial cutting activity involves removing the cord (C_3, one slice). After this cord has been removed, the next activity involves cutting through the box (C_3, one slice). To slice three sides of the box so that the lid can be lifted, the analyst would select the index value based on the number of strokes.

A_1 B_0 G_1 A_1 B_0 P_1 C_{10} A_1 B_0 P_1 A_0 =160 TMU

The criterion for selecting the index value to account for the initial placement of a knife is the same as discussed in the General Move chapter for Place. However, as a general rule, a P_1 will be sufficient. If the cut must be exact, P_3 will be appropriate.

Tool Use Examples for Cut

1. An operator picks up a knife from a workbench two steps away, makes one cut across the top of a cardboard box, and sets aside the knife on the workbench.

A_3 B_0 G_1 A_3 B_0 P_1 C_3 A_3 B_0 P_1 A_0

(3+1+3+1+3+3+1) X 10 = 150 TMU

2. During a sewing operation, a tailor cuts the thread from the machine before setting aside the finished garment. The scissors are held in the palm during the sewing operation.

A_0 B_0 G_0 A_1 B_0 P_1 C_1 A_0 B_0 P_0 A_0

(1+1+1) X 10 = 30 TMU

3. Following a soldering operation, an electronic component assembler must cut off the excess small-gauge wire from a terminal connection. The pliers are located within reach.

A_1 B_0 G_1 A_1 B_0 P_1 C_3 A_1 B_0 P_1 A_0

(1+1+1+1+3+1+1) X 10 = 90 TMU

4. An electrician working on transmission lines takes a pair of pliers from the tool belt and cuts off a piece of line. The nature of the line is such that two hands are needed to cut through the wire.

A_1 B_0 G_1 A_1 B_0 P_1 C_{10} A_1 B_0 P_1 A_0

(1+1+1+1+10+1+1) X 10 = 160 TMU

Surface Treat

Surface treat covers the activities aimed at cleaning material or particles from or applying a substance, coating, or finish to the surface of an object. There are many types of activites that may be included in the surface treat category, such as lubricating, painting, cleaning, polishing, gluing, coating, and sanding. However, the data found in Table 4.2 under surface treat cover only those general cleaning activities performed with a rag or cloth, an air hose, or a brush. Other kinds of surface treat activities, if encountered, may be treated as special tools (see Chapter 7) and supplementary index values developed for those particular activities.

The cleaning tools covered by the S parameter from Table 4.2 include:

1. *Air hose or nozzle* for blowing small particles or chips out of a hole or cavity or from a surface
2. *Brush* for brushing particles, chips, or other debris from an object or surface
3. *Rag or Cloth* for wiping light oil or a similar substance from a surface

Index values for these cleaning tools are based primarily on the amount of surface area cleaned. In most cases, the number of square feet cleaned determines the appropriate value. To analyze cleaning a small area such as a hole or cavity in a part, jig, or fixture with an air hose, the value S_6 (point or cavity) is appropriate; if more than one cavity is cleaned in this manner, frequency the value S_6 along with the P parameter and add an action distance (A) to account for the distance between cavities.

To brush clean a small object, an S_6 is appropriate. A small object refers to brushing a jig, fixture, or cavity. For example, air clean five holes with an air hose. The holes are > 2 in. (5 cm) apart.

$$A_1 \quad B_0 \quad G_1 \quad A_0 \quad B_0 \quad (P_1 \quad A_1 \quad S_6) \quad A_1 \quad B_0 \quad P_1 \quad A_0 \quad (5) = 440 \text{ TMU}$$

Tool Use Examples for Surface Treat

1. Before marking off a piece of sheet metal (4 ft.^{2}, 0.36 m^2) for a cutting operation, the operator takes a rag from his or her back pocket and wipes an oily film from the surface.

$$A_1 \quad B_0 \quad G_1 \quad A_1 \quad B_0 \quad P_1 \quad S_{32} \quad A_1 \quad B_0 \quad P_1 \quad A_0$$

$$(1+1+1+1+32+1+1) \times 10 = 380 \text{ TMU}$$

2. Following a sanding operation, an operator standing at a workbench picks up a brush located within reach and brushes the dust and chips from the working area (approximately 6 ft.^{2}, 0.54 m^2) and then sets aside the brush on the workbench.

$$A_1 \quad B_0 \quad G_1 \quad A_1 \quad B_0 \quad P_1 \quad S_{42} \quad A_1 \quad B_0 \quad P_1 \quad A_0$$

$$(1+1+1+1+42+1+1) \times 10 = 480 \text{ TMU}$$

3. Before assembling three components to a casting, the operator obtains an air hose (located within reach) and blows the small metal filings left from the previous machining operation out of three cavities. The distance between each cavity is > 2 in. (5 cm).

$$A_1 \quad B_0 \quad G_1 \quad A_0 \quad B_0 \quad (P_1 \quad A_1 \quad S_6) \quad A_1 \quad B_0 \quad P_1 \quad A_0 \quad (3)$$

$$[(1+1+6) \times (3)+1+1+1+1] \times 10 = 280 \text{ TMU}$$

Measure

Measure refers to the actions employed to determine a certain physical characteristic of an object by comparison with a standard measuring device.

Index values for the M parameter cover all actions necessary to place, align, adjust, and examine both the gauge and the object during the measuring activity. Therefore, the initial placement of the tool will normally be analyzed with a P_1. The data from Table 4.2 cover the following gauges.

M_{10} Profile Gauge

This value covers the use of an angle, radius, level, or screw-pitch gauge to compare the profile of the object to that of the gauge. The M_{10} value includes placing and adjusting the gauge to the object, plus the visual actions to compare the configuration of the object with that of the gauge.

M_{16} Fixed Scale

This parameter covers the use of a linear (12-in. (30-cm) ruler, yardstick, meter stick, etc.) or an angular (protractor) measuring device. The value M_{16} includes adjusting and readjusting the tool to two points and the time to read the actual dimension from the graduated scale.

M_{16} Calipers $\leqslant$ 12 in. (30 cm)

This parameter covers the use of Vernier calipers with a maximum measurement capacity of 12 in. (30 cm). The M_{16} value includes setting the caliper legs to the object dimension, locking the legs in place, and reading the Vernier scale to determine the measurement.

M_{24} Feeler Gauge

This parameter covers the use of a feeler gauge to measure the gap between two points. The M_{24} value includes fanning out the blades, reading and selecting the appropriate blade size, and positioning the blade to the gap to check for fit.

M_{32} Steel Tape $\leqslant$ 6 ft. (1.5 m)

This parameter covers the use of steel tape to measure the distance between two points. The M_{32} value includes pulling the tape from the reel, positioning the end of the tape, adjusting and readjusting the tape between the two points, the time to read the dimension from the scale, and finally pushing the tape back into the reel. This value is confined to the use of a steel tape from a fixed position, and includes *no* walking between the two points to adjust the tape.

M_{32}/M_{42}/M_{54} Micrometers $\leqslant$ 4 in. (10 cm)

These three index values cover the use of three different micrometers: M_{32} for measuring depth, M_{42} for measuring outside diameter, and M_{54} for measuring inside diameter. These values are based on micrometers designed for maximum dimensions of 4 in. (10 cm). The values include setting the micrometer to the part, adjusting the thimble for fit, locking the device, and finally reading the Vernier scale to determine the dimension.

Notice that for the above index values, all the placing and adjusting motions are included in the measure data (the M parameter). The result is that the initial placement of the measuring device is covered by each index value for M. For this reason, the place parameter prior to use tool will normally carry an index value of P_1 whenever the measure parameter is involved.

Tool Use Examples for Measure

1. Before welding two steel plates, a welder obtains a square and checks the angle between the plates to see that it is correct. The square (a profile gauge) is located three steps away on a workbench.

$A_6 \quad B_0 \quad G_1 \quad A_6 \quad B_0 \quad P_1 \quad M_{10} \quad A_6 \quad B_0 \quad P_1 \quad A_0$

$(6+1+6+1+10+6+1) \times 10 = 310$ TMU

2. Following a turning operation, a machinist checks the diameter of a small shaft with a micrometer. The micrometer is located on and returned to the workbench two steps away.

$A_3 \quad B_0 \quad G_1 \quad A_3 \quad B_0 \quad P_1 \quad M_{42} \quad A_3 \quad B_0 \quad P_1 \quad A_0$

$(3+1+3+1+42+3+1) \times 10 = 540$ TMU

Several supplementary values for various measuring devices have been developed which do not appear on the data card. They are as follows.

M_6 Snap Gauge

Measure with a snap gauge an outer diameter up to 2 in. (5 cm).

M_{10} Snap Gauge

Measure with a snap gauge an outer diameter up to 4 in. (10 cm).

M_{16} Plug Gauge

Measure with a plug gauge, GO + NOGO ends, up to 1 in. (2.5 cm).

M_{24} Thread Gauge

Measure with a thread gauge, a male or female screw up to 1 in. (2.5 cm).

M_{24} Vernier Depth Gauge

Measure with a Vernier depth gauge up to 6 in. (15 cm).

M_{42} Thread Gauge

Measure with a thread gauge, a male or female screw 1 to 2 in. (2.5-5 cm).

Record

Record covers the manual actions performed with a writing instrument or marking tool for the purpose of recording information. Three columns of data are found in Table 4.2 for the record parameter. The index values for write apply to the normal-size handwriting operations (script or print) performed with a pen, pencil, or other writing instrument. The mark data cover the use of such marking tools as a scribe, felt marker, or chalk, for the purpose of identifying or making a larger mark (1-3 in., 2.5-7.5 cm) on an object.

Write

These data are provided to cover the routine clerical activities encountered in many shop operations. These activities may include filling out time cards, writing out a part number, or writing brief instructions. Index values for the R parameter are selected primarily on the basis of the number of digits (letters or numerals) or the number of words written. Consider the values for writing the date (either in the form 07-04-80 or July 4, 1980) or writing one's signature as writing two words and assign an R_{16} for either.

Mark

This data applies to marking or identifying an object or container using a marking tool such as a scribe or felt marker. The index values for marking digits apply to printed characters (letters and numerals) of 1 to 3 in. in size. Other common marking values include making a check mark (R_1) and a scribing line (R_3).

The initial placement of a recording instrument before writing or marking usually occurs as a P_1. A possible exception may be the placement of a marking device prior to scribing a line. If the beginning point of the line is critical, a P_3 will be required to cover the necessary adjustment to place the tool accurately.

Tool Use Examples for Record

1. After finishing an assigned job, the operator picks up a clipboard and pencil (simo) from the workbench, fills out the completion date on the job card, and signs his name. He then returns the board and pencil to the workbench.

$$A_1\ B_0\ G_1\ A_1\ B_0\ \left(P_1\ A_0\ R_{16}\right)\ A_1\ B_0\ P_1\ A_0\ (2)$$

$$\left[(1+0+16)\ X\ (2)+1+1+1+1+1\right]\ X\ 10 = 390\ TMU$$

2. To order a part, a clerk takes a pencil from her shirt pocket and *writes* out a six-digit part number on the requisition form on her desk. She then clips the pencil back in her pocket.

$$A_1\ B_0\ G_1\ A_1\ B_0\ P_1\ R_{10}\ A_1\ B_0\ P_3\ A_0$$

$$(1+1+1+1+10+1+3)\ X\ 10 = 180\ TMU$$

3. Part of a packing operation involves identifying the components in the carton. This involves picking up a felt marker (within reach) and marking a six-digit number on the container.

A_1 B_0 G_1 A_1 B_0 P_1 R_{24} A_1 B_0 P_1 A_0

$(1 + 1 + 1 + 1 + 24 + 1 + 1) \times 10 = 300$ TMU

Think

Think refers to the use of the mental processes, particularly those involving visual perception. The think data in Table 4.2 are designed to cover only those types of reading and inspection activities that occur as a necessary part of a worker's job. While these operations usually occur internal to the manual work, and therefore have no effect on the duration of the work cycle, there are numerous occasions when these activities must be considered in the overall work content of the job. The analyst should exercise care in determining the extent to which these activities affect the total operation time.

Inspect

The data in this column apply to inspection work designed for making simple decisions regarding certain characteristics of the object under inspection. The activity involves first locating the inspection point (locations) and then making a quick yes-or-no decision concerning the existence of a defect. These mental processes presume that the inspector possesses a clear understanding of the characteristic being judged. In other words, the presence of any defect, such as a scratch, stain, scar, or color variance, is readily apparent to the inspector.

The index values for inspection refer to the number of inspection points examined on the object. For each point, a yes-or-no decision is made concerning the presence or absence of readily distinguishable defects.

These parameter values *do not* cover the manual handling of the object which might possibly occur during the inspection. Caution should be exercised in using these or any inspection values. In practical work situations, inspection time will rarely ever be limiting, but will usually occur during the manual handling of elements. In fact, work should be designed to maximize the number of inspections during manual activities whenever possible.

Along with inspecting a number of points, values are provided for activities of Touch for Heat (T_6), where the hand is moved to the object, moved over the surface of the object, and removed, and Feel for Defect (T_{10}), where the hand is moved to the object, moved over three surfaces of the object, and removed.

Read

To read is to locate and interpret characters or groups of characters. The data for Read are divided into two sections: Read "digits or single words" and Read "text of words".

The column "digits or single words" is to be used for reading technical data such as part numbers, codes, quantities, and dimensions from a blueprint. A digit is considered to be a letter, a number, or a special character. To index the T parameter, simply count the number of digits or single words read and choose the appropriate index number from the data card (Table 4.2).

The column "text of words" is to be used when analyzing situations in which the operator is required to read words arranged in sentences or paragraphs. These data are based on an average reading rate of 330 words per minute or 5.05 TMU per word. These index values may be applied to reading a set of instructions in a work order set or gathering general information from reading the tabular data or text on a blueprint.

Additional values which apply to more specific reading activities are also provided in Table 4.2.

T_3 Gauge

Use when a device is checked to see if the pointer is within a clearly marked tolerance range. *Example:* The pointer is in the range; the pressure is acceptable.

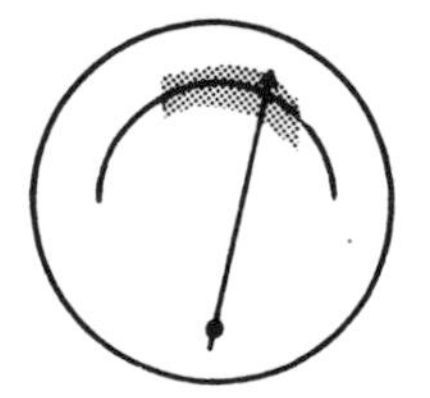

T_6 Scale Value

A specific quantity is read from a graduated scale such as a measuring stick or a temperature-pressure gauge. In the example below, the pressure is 38 psi.

T_6 Date/Time

The month, day, and year are read from a document or calendar; the time of day is read from a clock or wrist watch.

T_{10} Vernier Scale

Visually locate and read (only) an exact value from a micrometer, caliper or similar device. Does not contain time for placing and setting the device to an object.

T_{16} Table Value

A specific value is located and read from a table after scanning the table horizontally and vertically. *Example:* The correct machine setting is read from a feed-speed table.

Tool Use Examples for Think

1. During a testing operation, an electronics technician picks up a meter lead, places it on a terminal, and reads voltage off the meter scale. The lead is then put aside.

A_1 B_0 G_1 A_1 B_0 P_3 T_6 A_1 B_0 P_1 A_0

(1+1+1+3+6+1+1) X 10 = 140 TMU

2. Prior to starting a turning operation, an operator picks up a work order set and reads a paragraph which describes the method to be followed; contains an average of 30 words. The operator then places the set aside on the workbench.

A_1 B_0 G_1 A_1 B_0 P_0 T_{16} A_1 B_0 P_1 A_0

(1+1+1+16+1+1) X 10 = 210 TMU

5
The Equipment Handling Sequences

As stated in the introduction, the three sequence models covering the manual handling of objects constitute the basic MOST Work Measurement technique. These sequence models, General Move and Controlled Move in particular, can be used to measure handling of heavy objects, with lifting or moving equipment as well. However, for reasons of simplicity, special sequences were developed to cover this equipment handling. Because of the infrequent occurrence of these activities, sequence model parameter values were constructed from time intervals based on larger allowed deviations. For example, the Powered Crane and Truck Sequences were designed to use an index value multiplier of 100 instead of 10. This means that larger segments of work can be analyzed using fewer values.

The values appearing on the data cards for equipment handling are based on a representative sample of equipment found in industry. Therefore, the data may be valid for the majority of situations as they stand. However, before applying the data, it is suggested that individual parameter values be reviewed and adjusted to local methods if necessary. The primary considerations are the index values including crane and truck speeds. These values can of course be revised, using index value development procedures in the section, Application of MOST, in Chapter 7.

The three special MOST sequence models for moving heavy objects with material handling equipment are:

- Move with a manual crane
- Move with a powered crane
- Move with a truck

The Manual Crane Sequence

The Equipment Handling Sequences, as do the Manual Handling Sequences, indicate that a standard sequence of events must be considered when moving an object. One equipment handling sequence model, the Manual Crane Sequence Model, deals with the movement of objects using a manually traversed crane. The sequence model is appropriate for a crane which may resemble either a jib crane (Figure 5.1) or an overhead bridge crane (Figure 5.1), as long as the crane is moved laterally and longitudinally by hand, not under power.

As with the General Move Sequence, all manual operations can be identified with a certain sequence of events which repeats from cycle to cycle, regardless of the description, size, or name of the object being moved.

1. The operator moves to the crane (A).
2. The crane is transported empty to the location of the object to be moved (T).
3. The object is hooked up and freed from its surroundings (K, F).
4. The object is raised vertically using the crane (V).
5. The crane is moved, with the load, to the placement location (L).
6. The object is lowered vertically (V).
7. The object is placed in a fixture, container, etc., and thereafter unhooked from the crane (P).
8. The crane is transported empty to a rest position (T).
9. The operator returns to the original location (A).

Figure 5.1 Manually traversed cranes: jib crane (left) and overhead crane (right).

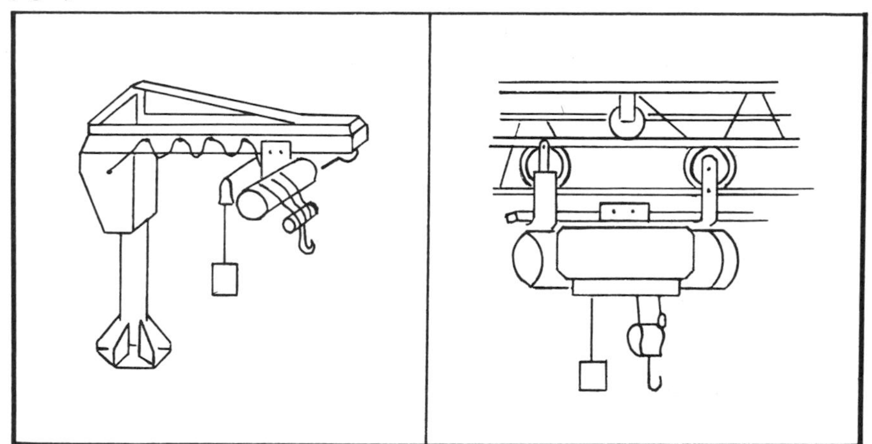

The Manual Crane Sequence Model

The movement of an object with a manual crane is described by the sequence model:

A T K F V L V P T A

where: A = Action distance
T = Transport empty crane
K = Hookup and unhook
F = Free object (from surroundings, pallet, fixture, etc.)
V = Vertical move
L = Loaded move
P = Place

Parameter Definitions

A Action Distance

This is defined in the section on General Move and is indexed by the distance (in steps) the operator walks to or from the crane.

T Transport Empty Crane

This refers to getting the empty crane and transporting it horizontally to the location of the object to be moved. Note that the movement is a result of the operator pulling or pushing the crane from one location to another. Vertical movement of the hook during the Transport empty parameter is an *internal* function.

K Hookup and Unhook

This includes both connecting and disconnecting the object from the holding device. The parameter begins at the point at which the transport of the empty crane ends and is complete when the object is fastened to the crane hook, sling, etc. The parameter also includes time to remove the holding device. Note that getting the hook or slings to the workplace will be analyzed separately with a General Move Sequence.

F Free Object

This refers to the actions necessary to work the object free from its surroundings (e.g., container or fixture) and raise the object, at a low speed, 2 to 3 in. (5-8 cm). This parameter is to include all actions necessary to position the load so that the next activity will be an unrestricted vertical move.

V Vertical Move

This is the raising or lowering of the object at high speed following the F and L parameters. The hook is raised after the object is freed and lowered after the loaded crane is moved to the placement location. Note that if the hook is raised or lowered during the transportation of the crane, the time is covered by the T or L parameters.

L Loaded Move

This refers to the horizontal movement of the object with the crane. Note that the movement with a manual crane is a result of the operator pulling or pushing the crane from one location to another.

P Place

This covers the actions in lowering the object the last 2 to 3 in. (5-8 cm) at low speed and placing the object in the desired location. Index values are based on the degree of difficulty affecting placement, as follows.

1. No alignment: The object is simply lowered into position without any additional manual guidance from the operator. *Example:* Lower a small casting and place it by itself on a pallet.
2. Align with one hand: While lowering the last 2 to 3 in. (5-8 cm), the operator reaches out with one hand and steers or swings the load into position.
3. Align with two hands: During the placement activity, the operator must release the controls and steer or swing the object into position using two hands.
4. Align and place with one adjustment: To position an object, the operator must steer or swing the object into position, in addition to making one directional adjustment (longitudinally, laterally, or vertically).
5. Align and place with several adjustments: To position an object, an operator must steer or swing the object into position, in addition to making several directional adjustments (longitudinally, laterally, and/or vertically).
6. Align and place with several adjustments, in addition to care in handling or applying pressure: To position an object, an operator must steer or swing the object into position, in addition to making several directional adjustments (longitudinally, laterally, and/or vertically). A pause or hesitation must also be observed at the point of placement to indicate the application of heavy pressure required to seat the object, or an obvious slow motion is observed in placing the object carefully.

Figure 5.2 is presented to illustrate the sequence of events that occurs when an object is moved with a manual crane.

Use of the Manual Crane Data Card

The data card (Table 5.1) is divided into seven columns. Index values are selected either by the distance involved (the A, T, L, and V parameters) or by the holding device used or difficulty involved in moving an object (the F and P parameters).

Sequence Model Indexing

A Action Distance

Choose the index value by the distance the operator walks to get to or move away from the crane.

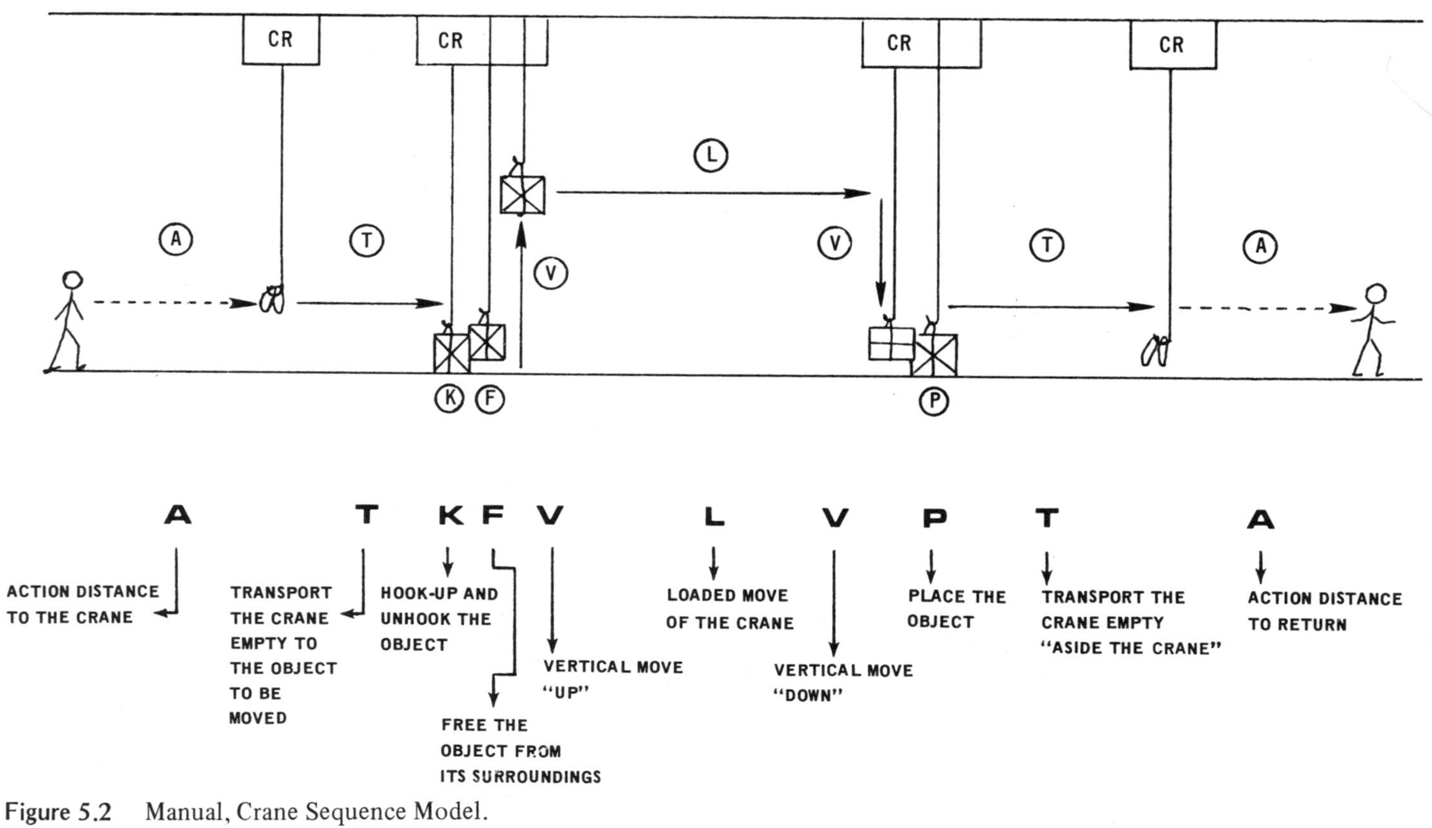

Figure 5.2 Manual, Crane Sequence Model.

Table 5.1 Manual Crane Data Card*

	ATKFVLVPTA	MANUAL CRANE SEQUENCE MULTIPLIER 10						
	A	T	L	K	F	V	P	
INDEX	ACTION DISTANCE STEPS	TRANSPORTATION UP TO 2 TON FEET (M) EMPTY	LOADED	HOOK-UP AND UNHOOK	FREE OBJECT	VERTICAL MOVE IN (CM)	PLACE OBJECT	INDEX
3	1-2	-	-		WITHOUT DIRECTION CHANGE	9 (20)	WITHOUT DIRECTION CHANGE	3
6	4	-	-		WITH SINGLE DIRECTION CHANGE	15 (40)	ALIGN WITH ONE HAND	6
10	7	5 (1.5)	5 (1.5)		WITH DOUBLE DIRECTION CHANGE	30 (75)	ALIGN WITH TWO HANDS	10
16	10	13 (4)	12 (3.5)		WITH ONE OR MORE DIRECTION CHANGES, CARE IN HANDLING OR APPLY PRESSURE	45 (115)	ALIGN AND PLACE WITH ONE ADJUSTMENT	16
24	15	20 (6)	18 (5.5)	SINGLE OR DOUBLE HOOK		60 (150)	ALIGN AND PLACE WITH SEVERAL ADJUSTMENTS	24
32	20	30 (9)	26 (8)	SLING			ALIGN AND PLACE WITH SEVERAL ADJUSTMENTS & APPLY PRESSURE	32
42	26	40 (12)	35 (10)					42
54	33	50 (15)	45 (13)					54

*Values are read "up to and including."

T Transport Empty

Select the proper index value by the distance (feet, meters) the operator moves the empty crane to or from the object moved. *Note:* The values for up to 2 tons include the use of ¼-ton, ½-ton, 1-ton, and 2-ton cranes.

L Loaded Move

Select the proper index value by the distance (feet, meters) the operator moves the loaded crane.

K Hookup and Unhook

Choose the proper index value by the holding device used.

F Free Object

Choose the proper index value by the difficulty involved in freeing the object, in other words, raising the object 2 to 3 in. (5-8 cm) and positioning such that the next action will be an unobstructed vertical move.

V Vertical Move

Select the proper index value by the distance (inches, centimeters) the object is raised or lowered.

P Place

Choose the proper index value by the difficulty involved in lowering the object the last 2 to 3 in. (5–8 cm) and placing it in the desired location.

Like the General Move, Controlled Move, and Tool Use Sequences, the index numbers are added for one sequence model and the total is multiplied by 10 to convert to TMU.

Manual Crane Data Card Backup Verification The data provided on this data card are to be treated as sample data only. The methods represented on this card must be verified and the vertical speeds (process time) must be validated for the particular cranes in question.

Methods to be verified are the hook and unhook (K) and place (P) subactivities. Backup data for methods other than specified on the data card can be developed and placed on the data card according to the procedures outlined in Chapter 7. Equipment data to be verified and validated are loaded and unloaded transportation speeds and vertical speeds (T, L, V).

The transportation time per traveled distance and the corresponding index values can be calculated using the following formula:

$t = c + (s \times n)$

where t = time in TMU

n = distance variable in number of feet moved

c = fixed manual time (TMU), including grasping of control and crane acceleration and deceleration times
s = crane horizontal speed (TMU/ft.)

Manual Crane Sequence Examples

1. A machine operator walks 10 ft. (3 m) to a crane and manually transports it to a fixture (66 lb., 30 kg) located 7 ft. (2 m) away. The fixture, which is lying by itself on a pallet, is hooked up to the crane with a single hook and moved 14 ft. (4.5 m) to a workbench 3 ft. (1 m) higher than the pallet. The fixture is then lowered 4 in. (10 cm) and placed on top of the workbench. The operator transports the empty crane 3 ft. (1 m) away and returns to the workbench.

$A_6 \quad T_{16} \quad K_{24} \quad F_3 \quad V_{16} \quad L_{24} \quad V_3 \quad P_3 \quad T_{10} \quad A_3$

$(6+16+24+3+16+24+3+3+10+3) \times 10 = 1080$ TMU

2. The activity involved in exchanging the workpiece in a 3-jaw chuck of an engine lathe requires the use of a jib crane. The operator first gets the jib crane from two steps away and transports it back to the machine where he or she hooks up the 300-lb workpiece with a sling. He or she raises the crane 6 in. (15 cm) and moves the load 16 ft. (5 m) away and then lowers the crane 3 ft. (1 m) to place the workpiece on a pallet. From another pallet located 6 ft. (2 m) from the first, the operator gets a new workpiece, moves it back to the machine (22 ft. 7 m) and places it in the chuck, then puts the crane aside, two steps away, and returns.

$A_3 \quad T_{10} \quad K_{32} \quad F_{16} \quad V_3 \quad L_{24} \quad V_{16} \quad P_3 \quad T_0 \quad A_0 = 1070$ TMU

$A_0 \quad T_{16} \quad K_{32} \quad F_3 \quad V_{16} \quad L_{32} \quad V_3 \quad P_{32} \quad T_{10} \quad A_3 = 1470$ TMU

2540 TMU TOTAL

The Powered Crane Sequence

The second of the three Equipment Handling Sequence Models deals with moving an object with the assistance of a power traversed crane. The sequence model is appropriate for a crane which may resemble an overhead, pendant-operated bridge crane (Fig. 5.3).

The Powered Crane Sequence Model is appropriate for cranes that move the load laterally and longitudinally under power.

The use of a pendant bridge crane consists of the following activities.

1. The operator walks to the control panel (A).
2. The operator grasps the controls, elevates the crane hook, moves the crane

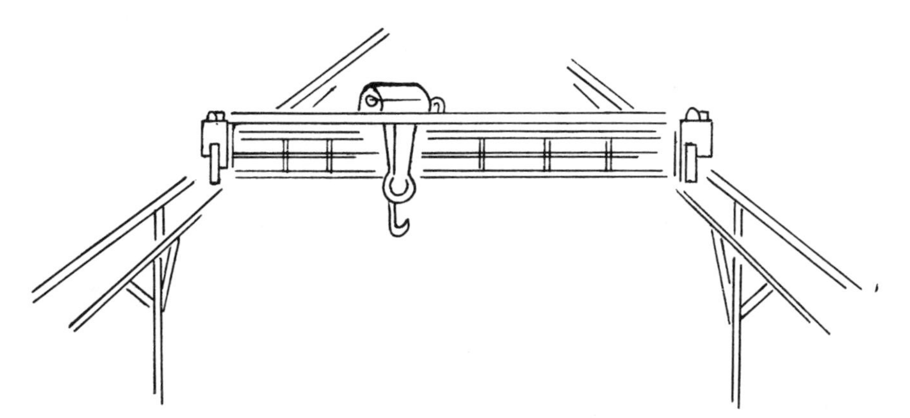

Figure 5.3 Overhead bridge crane.

so the hook reaches the position for coupling, and then releases the controls (T).

3. The object is fastened either directly with the crane hook or with a sling, chain, etc. The operator then grasps the controls and elevates the crane hook to the correct position for hooking and then adjusts the controls so that the chain or other holding device is tight and secure (K). (The holding device is subsequently removed from the object.)
4. The crane hook, with the object is freed from its surroundings and elevated so the object can be moved. The object is then moved horizontally to the desired location (T).
5. The object is lowered and placed in the desired location (P).
6. The empty crane is moved aside (T).
7. The operator returns to the starting point after moving the crane aside (A).

The Powered Crane Sequence Model

The above activities can be described by the following sequence model:

A T K T P T A

where A = Action distance
T = Transport
K = Hookup and unhook
P = Place

If the local conditions call for a permanent bridge crane operator, only the K, T, and P parameters are needed for analysis purposes. Note that the index values for the Bridge Crane Sequence are multiplied by 100 for conversion to TMU.

Powered Crane Data Card Backup Information

It is recognized that there are many manufacturers, models, capacities, etc., of cranes available. As a result, the information presented on the powered crane data card, as determined by the backup data outlined below, should be treated as sample information. The method must be verified and the process times must be validated to fit a company's particular equipment.

The backup data below are provided as a key to the kinds of considerations that should be made when validating or developing appropriate index values for particular situations or cranes.

Basic Data for Powered Crane Parameter Values

The times listed below were derived from a series of time studies.

1. Longitudinal traverse: low speed = 122 TMU/ft. (400 TMU/m); high speed = 15.3 TMU/ft. (50 TMU/m).
2. Vertical move: low speed = 30 TMU/in. (1200 TMU/m); high speed = 7.6 TMU/in. (300 TMU/m).
3. Lateral traverse = 152.5 TMU/ft. (500 TMU/m).
4. Change direction using a control panel button: 40 TMU.
5. Lateral tranportation frequency = 10%.

Parameter Definitions

A Action Distance

This value is the horizontal distance the operator travels to or from the bridge crane control panel. Note that parameter values are different from those in General Move because of the larger index value multiplier.

T Transport

This parameter refers to movement of the crane with or without a load. The activities listed below were included in the T parameter provided on the data card and must be verified and validated to fit particular situations and/or cranes.

Unloaded	TMU	Loaded	TMU
Grasp control panel and depress button	40	Grasp control panel and depress button	40

Unloaded	TMU	Loaded	TMU
Raise 3.3 ft. (1 m) (high speed)	300	Raise 1.5 ft. (0.5 m) (creep speed)	600
Acceleration and retardation	200	Raise 5 ft. (1.5 m) (high speed)	450
Change direction	40	Acceleration and retardation	500
Long, traverse (variable)	---	Change direction	40
Change direction	40	Long traverse (variable)	---
Lower 3.3 ft. (1 m) (high speed)	300	Change direction	40
Total	920		1670

Transport (T) Constant Development Since the T parameter is appropriate for both loaded and unloaded movement of the crane, the weighted T constant was developed to include:

- Unloaded Move, to move the crane to the object or aside with a frequency of 60% — 920 TMU X 0.60 = 552 TMU
- Loaded Move, to move the crane with the object, 40% — 1670 TMU X 0.40 = 668 TMU

Constant Total Time = 1220 TMU

Note that the T constant includes the freeing of the object using creep speed and average vertical moves before or after the horizontal movement. Vertical moves made during the horizontal movement are internal.

The average horizontal speed is determined by weighting the longitudinal and lateral speeds according to work area observations (lateral speed frequency). Only the high speed value is to be used in the calculation, as the low speed is included in the acceleration and deceleration times for the prevention of object swinging.

The weighted horizontal speed includes:

- Lateral speed — 152.5 TMU/ft. X 0.10 = 15.3 TMU/ft.
- Longitudinal speed — 15.3 TMU/ft. X 0.90 = 13.8 TMU/ft.

Horizontal Speed = 29.1 TMU/ft.

The transport (T) time and the corresponding index values are based on the following formula:

$t = 1220 + (29.1 \times n)$

where t = time in TMU
1220 = move constant (TMU)
29.1 = horizontal transportation speed in TMU/ft.
n = number of feet moved

K Hookup and Unhook

This parameter includes the activities involved in both connecting and disconnecting the object from the crane. The parameter begins when the hook has been transported close to the hooking position and is completed when the holding device has been disconnected from the object. Adjustments of the crane, for hooking and for tightening and securing the holding device, are included. Hooking and unhooking refers to fastening chains, slings, or other holding devices to the crane hook or to the object.

Getting and moving aside chains, sling, etc., and fastening them to the object (or to the crane hook) initially are not covered by the K parameter. Such activities are analyzed with General Move and Controlled Move Sequences.

Hookup and unhook index values are chosen by the type of holding device used.

P Place

This parameter involves all actions necessary to lower the object with a combination of high speed and creep speed and to place the object in the desired location. Index values are based on the following data.

1. With or without single direction change		
Lower 20 in. (50 cm) (high speed)		150
Lower 2 to 3 in. (5-8 cm) (low speed)		75
Change direction		40
½ longitudinal traverse 4 in. (10 cm)		20
½ lateral traverse 4 in. (10 cm)		25
	Total	310 TMU
2. With double direction change		
Lower 5 ft. (1.5 m) (high speed)		450
Lower 20 in. (50 cm) (low speed)		600
4 × change direction		160
Longitudinal traverse 20 in. (50 cm)		200
Lateral traverse 20 in. (50 cm)		250
	Total	1660 TMU
3. With several direction changes		
Lower 6.6 ft. (2 m) (high speed)		600
Lower 20 in. (50 cm) (low speed)		600
8 × change direction		320

Longitudinal traverse 3.3 ft. (1 m) (low speed)		400
Lateral traverse 3.3 ft. (1 m)		500
	Total	2420 TMU

Figure 5.4 is presented to illustrate the sequence of events that occurs when an object is moved with a power traversed crane. *Note:* The data in Table 5.2 should be treated as *sample data* and must be validated prior to their use.

Use of the Powered Crane Data Card

The data card (Table 5.2) is divided into four columns. Index values are selected, by the distance involved (A and T), by the holding device used (K), or by the difficulty involved (P) in placing the object.

Sequence Model Indexing

A Action Distance

Choose the index value by the distance the operator walks to get to or move away from the crane.

T Transport

Choose the index value by the distance the crane is moved horizontally, either loaded or unloaded. All vertical distances are included in the (T) values; separate vertical analyses are not necessary.

K Hook-up and Unhook

Choose the proper index value by the holding device used.

P Place

Choose the proper index value by the difficulty involved in lowering the object the last 2 to 3 in. (5–8 cm) and placing it in the desired location.

Once the sequence model is indexed, the index numbers should be added and the total multiplied by 100 to convert to TMU. *Example:* An operator walks 90 ft. (27 m) to a powered crane control panel and transports the crane to a part 25 ft. (8 m) away. The part is connected to the crane with one hook and a sling and transported 2 ft. (0.5 m) where it is placed with a double change of direction. The operator then moves the crane 9 ft. (3 m) out of the way and walks back to the part.

A_6 T_{16} K_{24} T_{10} P_{16} T_{16} A_1 8900 TMU

$(6 + 16 + 24 + 10 + 16 + 16 + 1) \times 100 = 8900$ TMU

The Truck Sequence

The Truck Sequence is primarily concerned with the horizontal transportation of material from one location to another using a "wheeled device." Equipment

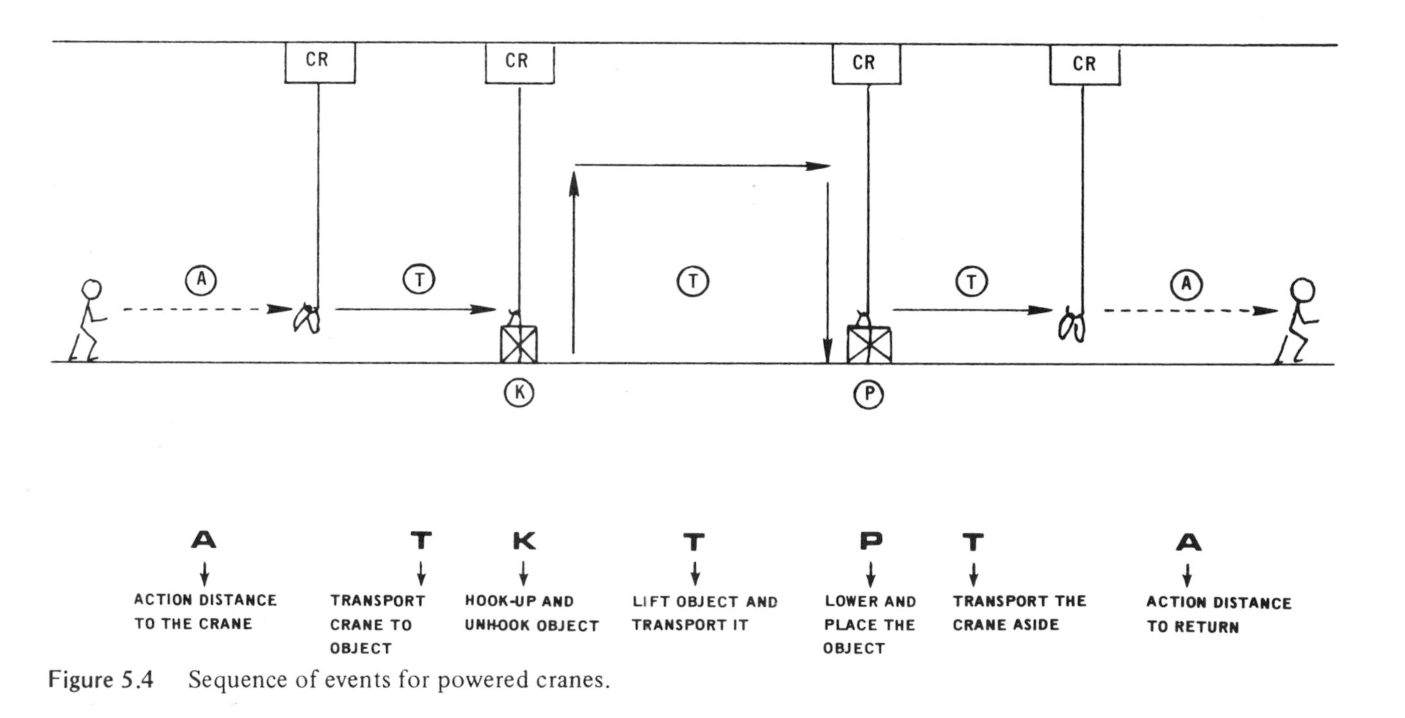

Figure 5.4 Sequence of events for powered cranes.

Table 5.2 Powered Crane Data Card*

m	ATKTPTA		POWERED CRANE SEQUENCE 10 - 20 TON CRANE MULTIPLIER 100		
INDEX	A ACTION DISTANCE FEET (M)	T TRANSPORT FEET (M)	K HOOK-UP AND UNHOOK HOLDING DEVICE	P PLACE DIFFICULTY	INDEX
1	24 (7)				1
3	60 (18)			WITHOUT OR WITH SINGLE CHANGE OF DIRECTION	3
6	120 (36)		SINGLE HOOK OR ELECTROMAGNET		6
10	190 (60)	2 (0.5)			10
16	300 (90)	25 (8)		WITH DOUBLE CHANGE OF DIRECTION	16
24	420 (130)	50 (16)	1 HOOK PLUS SLINGS OR CHAINS	WITH SEVERAL CHANGES OF DIRECTION	24
32	550 (170)	80 (25)	2 HOOKS PLUS SLINGS OR CHAINS		32

*Values are read "up to and including."

covered by this sequence falls within two general categories: trucks operated from a "riding" position and those requiring "walking" (see Fig. 5.5).

Riding trucks: fork lift truck, high stacker

Walking trucks: hand truck (two to four wheels), low lift pallet truck, stacker

Transportation of material with trucks consists of the following activities:

1. The operator walks to the truck (A).
2. The operator takes a seat (if riding) and starts the truck (S).
3. The truck is driven or transported to the material (T).
4. The material is picked up or loaded by the fork or lifting attachment (L).
5. The material is transported to another location (T).
6. The material is unloaded (L).
7. The truck is driven to another area and parked (T).
8. The operator returns to the original (or another) location (A).

The Truck Sequence Model

The above activities are described by the following sequence model:

A S T L T L T A

where A = Action distance
S = Start and park
T = Transport
L = Load or unload

To more easily see the sequence of events that occurs in moving an object with a wheeled truck, follow the sequence model and the events as pictured in Figure 5.6.

Parameter Definitions

A Action Distance

This is the distance walked by the operator to or from the truck. Note that these parameter values are different from those in General Move.

S Start and Park

This value includes the actions to prepare the truck for moving plus the parking activity following the final transport. For the *riding truck* these activities include climbing in and out of the seat, starting and stopping the engine, releasing and engaging the hand brake, etc. For the *walking truck* these activities include taking hold of the handle, starting and stopping the power, if applicable, tilting the body or handle.

T Transport

This parameter applies to the movement of the truck with or without a load.

RIDING TRUCKS

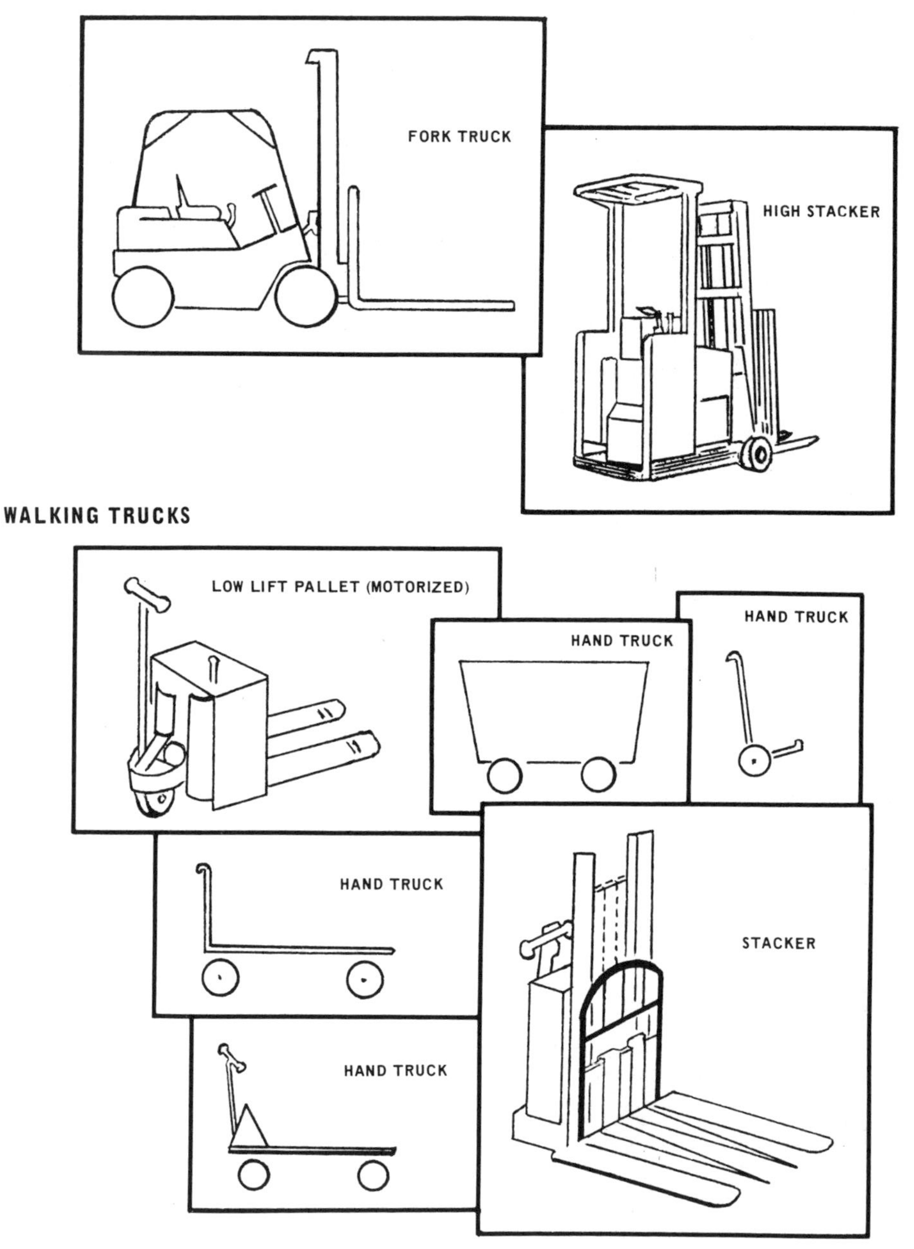

Figure 5.5

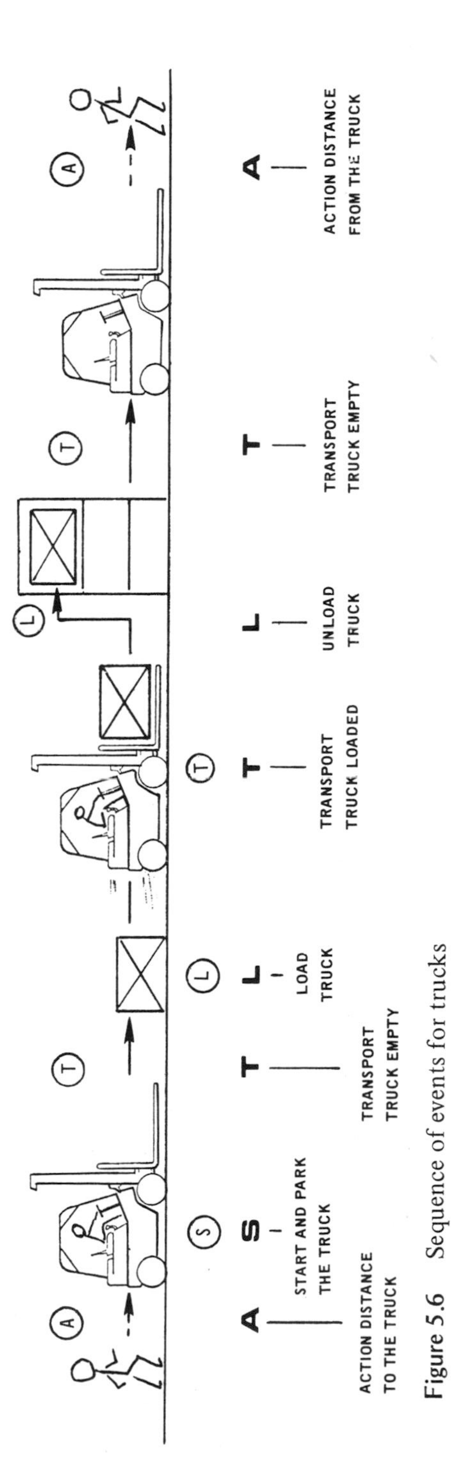

Figure 5.6 Sequence of events for trucks

L Load or Unload

This involves either picking up the material at the original location or placing the material at the destination using the forks or other lifting attachments. If a hand truck is used, however, loading and unloading is analyzed with the General Move Sequence. If such is the case, the L parameters of the Truck Sequence Model will carry a zero index value.

Use of the Truck Data Card

The data card (Table 5.3) is divided into four main sections, each representing a parameter as defined above. Index values are selected by the distance involved (A and T), by the type of truck used (S), or by the location of the object (L).

Sequence Model Indexing

A Action Distance

Choose the index value by the distance the operator walks to get to or move away from the truck.

S Start and Park

Choose the index value by the type of truck used: S_3, walking truck; S_6 riding truck.

T Transport

First, choose the correct column by the general truck type (riding or walking) and then the specific kind of truck (forklift, high stacker, stacker, low-lift pallet truck or hand truck. Next, select the index value based on the distance (in feet or meters) that the truck is transported.

L Load *or* Unload

Choose the correct index value by the location of the object when loading or unloading: L_6, used when loading or unloading an object either from or to the "floor"; L_{10}, used when loading or unloading an object either from or to a "pallet rack" (above the floor).

Again, due to the vast number of manufacturers and seemingly infinite configuration of trucks, the data provided in Table 5.3 should be treated as *sample* information. Before its use in establishing a labor standard, the method must be verified and the process time *must be validated* to a company's particular trucks and conditions. Once index values are selected from the data card, they are placed on the Truck Sequence Model, added, and multiplied by 100 to convert to TMU.

Table 5.3 Truck Data Card*

ASTLTLTA	TRUCK SEQUENCE MULTIPLIER 100								
	A	S	T					L	
			TRANSPORT WITH OR WITHOUT LOAD - FEET (M)						
			RIDING		WALKING				
INDEX	ACTION DISTANCE FEET (M)	START AND PARK	FORK LIFT TRUCK	HIGH STACKER	STACKER	LOW LIFT PALLET TRUCK	HAND TRUCK	LOAD OR UNLOAD	INDEX
1	24 (7)		27 (8)	21 (6)	11 (3)	14 (4)	24 (7)		1
3	60 (18)	WALKING TRUCK	67 (20)	50 (15)	27 (8)	34 (10)	50 (15)		3
6	120 (36)	RIDING TRUCK	132 (40)	100 (30)	50 (15)	67 (20)	100 (30)	FLOOR	6
10	190 (60)		198 (60)	165 (50)	83 (25)	100 (30)	165 (50)	PALLET RACK	10
16	300 (90)		329 (100)	247 (75)	116 (35)	165 (50)	264 (80)		16
24	420 (130)		460 (140)	362 (110)	182 (55)				24

*Values are read "up to and including."

Truck Sequence Examples

1. An operator walks 120 ft. (36 m) to a forklift truck, climbs into the seat, and starts the engine. It is driven 12 ft. (4 m), where a pallet is picked from the floor and then transported 75 ft. (23 m) and placed in a pallet rack. The truck is then parked 30 ft. (9 m) away, and the operator returns 60 ft. (18 m) to the work place.

$A_6 \quad S_6 \quad T_1 \quad L_6 \quad T_6 \quad L_{10} \quad T_3 \quad A_3$

(6 + 6 + 1 + 6 + 6 + 10 + 3 + 3) X 100 = 4100 TMU

2. An operator walks six steps to a bench, picks up a heavy object, and places it on a low four-wheeled hand truck five steps away. After this, the operator takes the handle of the truck (four steps away) and transports the object 36 ft. (11 m). The truck is parked, and the operator walks 150 ft. (46 m) to another work area.

$A_{10} \quad B_0 \quad G_3 \quad A_{10} \quad B_6 \quad P_1 \quad A_0$

(10 + 3 + 10 + 6 + 1) X 10 = 300 TMU

$A_1 \quad S_3 \quad T_0 \quad L_0 \quad T_3 \quad L_0 \quad T_0 \quad A_{10}$

(1 + 3 + 3 + 10) X 100 = 1700 TMU

2000 TMU TOTAL

3. An operator walks 15 ft. (5 m) to a low lift pallet truck, starts it, and transports it 21 ft. (6 m) to a pallet located on the floor. The pallet is loaded on the truck and then transported 100 ft. (30 m) to a warehouse where it is then placed on the floor. The operator then transports the truck 89 ft. (27 m), parks it, and walks 30 ft. (9 m) to a workbench.

$A_1 \quad S_3 \quad T_3 \quad L_6 \quad T_{10} \quad L_6 \quad T_{10} \quad A_3$

(1 + 3 + 3 + 6 + 10 + 6 + 10 + 3) X 100 = 4200 TMU

6
MOST for Clerical Operations

The emergence in the 1940s and 1950s of predetermined motion time systems, especially those that focus on the clerical area, provided management personnel with the tool to determine the time needed to perform certain tasks, with minimal disruption in the office. However, the analysis time consumed by the detailed systems of that time, and the considerable amount of documentation required, resulted in a hesitation to use such work measurement systems. Also, clerical operations contained wide variations in the methods used to perform them, as little methods engineering time focused on clerical operations. These factors led to a predominant use of the stopwatch over the predetermined motion time systems as the best way to tackle the clerical work measurement task.

There have been many improvements in work measurement techniques during the 1960s and 1970s, with MOST being in the forefront. MOST Clerical Systems, unlike other clerical predetermined motion systems, is quickly learned and implemented. Its methods sensitivity encouraged methods engineers. By applying the MOST Clerical Systems, the managerial staff acquires accurate data and the standards needed to produce personnel tables, performance charts, and other meaningful management documents.

MOST Clerical Systems is a variation of the MOST Work Measurement System for general production. Although it is applied using the same sequence model and analysis format as the general production version, there are additions to the General and Controlled Move data cards. A new Equipment Use data card has also been added specifically to handle clerical operations. MOST Clerical Systems provides the same advantages for clerical work measurement as the general production version provides for industrial applications, and produces equivalent results.

Table 6.1 Manual Clerical Sequence Models

ACTIVITY	SEQUENCE MODEL	SUB-ACTIVITIES
GENERAL MOVE	ABGABPA	A - ACTION DISTANCE B - BODY MOTION G - GAIN CONTROL P - PLACE
CONTROLLED MOVE	ABGMXIA	M - MOVE CONTROLLED X - PROCESS TIME I - ALIGN
EQUIPMENT USE	ABGABP ABPA	H - LETTER / PAPER EQUIPMENT HANDLING T - THINK R - RECORD K - CALCULATE W - TYPEWRITE
TOOL USE	ABGABP ABPA	F - FASTEN L - LOOSEN C - CUT M - MEASURE

MOST Clerical Systems is based on three activity sequence models. They are the General Move, Controlled Move, and Tool/Equipment Use (see Table 6.1). Using the definitions provided in a separate text, the analyst indexes these sequence models in the same manner as previously mentioned in Chapters 2, 3, and 4.

The General Move

The Clerical General Move data card is based on the same theory as the general production version. The sequence model deals with spatial displacement of an object under manual control, in which the object's path is not restricted. Because of the nature of the clerical operations, additional values were added to the General Move data card. For example, collect and jog paper (G_6) includes the time needed for an operator to collect or gather several sheets of paper and jog them together, a common clerical activity. It is this type of activity that has been taken into consideration and provided for in the General Move data card of the MOST Clerical Systems.

An example of collect and jog: An operator collects several sheets of paper and jogs them twice and asides them to the "out" basket. The sequence model for analyzing this operation is as follows:

$A_1\ B_0\ G_6\ A_1\ B_0\ P_1\ A_0$

$(1+6+1+1) \times 10 = 90$ TMU

Another example of a value added to the clerical General Move data card is Gain Control with Intermediate moves (G_6). The value covers several moves or repositionings of the fingers or hands to gain complete control of the object.

An example of G_6 intermediate moves: The clerk prepares an envelope for mailing by removing a label from an adhesive pad (with intermediate moves) and places it on the envelope (with intermediate moves). The following sequence model reflects the above activity.

$A_1\ B_0\ G_6\ A_1\ B_0\ P_6\ A_0$

$(1+6+1+6) \times 10 = 140$ TMU

The Controlled Move

Like the General Move, the Clerical Controlled Move data card has been altered to include several additional values that provide for various clerical controlled move operations. The Controlled Move Sequence Model describes the manual displacement of an object over a "controlled" path. That is, the movement of an object is restricted in at least one direction by contact with or attachment to another object.

Many clerical examples of controlled moves occur when handling paper. Therefore, the Clerical Controlled Move data card provides additional information for these activities in the Move Controlled (M) parameter under the general heading of paper handling operations. This parameter applies to handling paper to change its shape, direction, or position (i.e., interleaf paper or unfold paper). For example, interleaf paper covers the action of lifting the sheet(s) of paper with one hand,while reaching and grasping a divider sheet, carbon paper, etc., with the other hand and inserting it beneath the sheet that was lifted.

An example of the interleaf paper parameter: An operator interleafs four sheets of blue divider among the first four sections of a report. The MOST analysis to interleaf all four sheets would be

$A_1\ B_0\ G_1\ M_6\ X_0\ I_0\ A_0\quad 4$

$(1+1+6) \times 4 = 32 \times 10 = 320$ TMU

Equipment Use

The major difference between the Clerical and General Production MOST Work Measurement Systems is the addition of new elements covering equipment use. An additional data card is used to analyze common clerical activities including common pieces of office equipment. The Equipment Use Sequence

Model shows the same format as the Tool Use Sequence Model and it is also mentally divided into five sections.

GET EQUIPMENT	PLACE EQUIPMENT	USE EQUIPMENT	ASIDE EQUIPMENT	RETURN
ABG	ABP		ABP	A

The open space or gap in the sequence model provides for the insertion of one of the following Equipment Use parameters:

where H = letter/paper equipment handling
T = think/read
W = typewrite
R = record
K = calculate

The Equipment Use data cards contain values for clerical functions such as stapling, sealing, stamping, typing, writing, and calculating. To apply the information appearing on these data cards, the user follows similar procedures to those outlined previously in Chapter 4. As an example, the subactivity typewrite, which appears on the Equipment Use data card, refers to the use of fingers and hands performing multiple Control and General Moves to type words, sentences, letters, headings, and so on. The index value chosen from the data card is based primarily on the number of characters typed or the functions performed on the typewriter. So, if, for example, a word processor inserts a sheet of paper into the platen in preparation for typing a letter, the appropriate value for this insertion would be a W_{24}. The sequence model for this insertion reads as follows.

$$A_1 \; B_0 \; G_1 \; A_1 \; B_0 \; P_0 \; W_{24} \; A_0 \; B_0 \; P_0 \; A_0$$

$(1+1+1+24) \times 10 = 270$ TMU

The same operator then types a 14-word letter with two carriage returns. The sequence model for typing the 14 words will read as follows:

$$A_0 \; B_0 \; G_0 \; A_0 \; B_0 \; P_0 \; W_{42} \; A_0 \; B_0 \; P_0 \; A_0$$

$42 \times 10 = 420$ TMU

The two carriage returns would be analyzed

$$A_0 \; B_0 \; G_0 \; M_1 \; X_0 \; I_0 \; A_0 \quad 2$$

$1 \times 2 = 2 \times 10 = 20$ TMU

Total time to perform the paper insertion plus typing the 14 words with the two carriage returns is equal to 710 TMU which represents a 100% performance level and includes *no* allowances.

Tool Use

During a clerical operation there may be a need to use one of several common hand tools (knife, scissors, ruler, brush, etc.). Values for the use of these tools are found on the Tool Use data card. Clerical Tool Use application procedures are identical to those stated in Chapter 4. Selection of the appropriate index value will be based on the tool used or, in some cases, the type and number of motions performed. For example, a secretary picks up a bottle of correction fluid and with six finger actions, unscrews the cap and places the bottle on the desk. The analyses for that operation are as follows.

$A_1\ B_0\ G_1\ A_1\ B_0\ P_1\ L_{10}\ A_1\ B_0\ P_1\ A_0$

$(1+1+1+1+10+1+1) \times 10 = 160$ TMU

Application of the MOST Clerical System

Often, in a clerical environment, the jobs performed by an operator vary widely. Clerical work is very seldom "repetitive." For example, a secretary may type three letters, one a 21-line (½-page) letter, the second 40-line (full-page) letter, and the third a 64-line (1½-page) letter. Although this example only reflects one type of variation, length, other variations such as margin size, tabular typing, and spacing, could be performed and could be numerous. The MOST Clerical Systems is capable of handling these variations because of the high application level, ease of analysis, and sound statistical principles upon which MOST Work Measurement Systems are based. Being an easy system to use and understand, as well as being method sensitive, MOST Clerical Systems provides the analyst with the tool for comparing the effects of various methods of performing clerical activities, and, in fact, "engineers" workplaces and methods to allow the clerical worker to work more efficiently and more productively. Also available within the MOST Clerical Application Systems is the statistical framework which makes analysis of variable situations possible. The clerical analyst might find any or all the approaches to the calculation of final time standards (Chap. 8) appropriate. Proper application of these approaches produces consistent and well-documented standard times. The system also allows a choice of standard calculation formats based on the type and variability of the work being analyzed.

MOST Clerical Computer Systems

Finally, all the advantages of the manual system are incorporated and enhanced in the MOST Clerical Computer Systems. Generation of standard times can occur 20 to 40% faster than with the manual system. By using the MOST Clerical Computer System, the analyst can create MOST analyses, suboperation data, and operation data, document the workplace, calculate performance, and establish a cost center personnel chart. *Note:* A glossary of terms and a more detailed description of the MOST Computer Systems appears in Chapter 9.

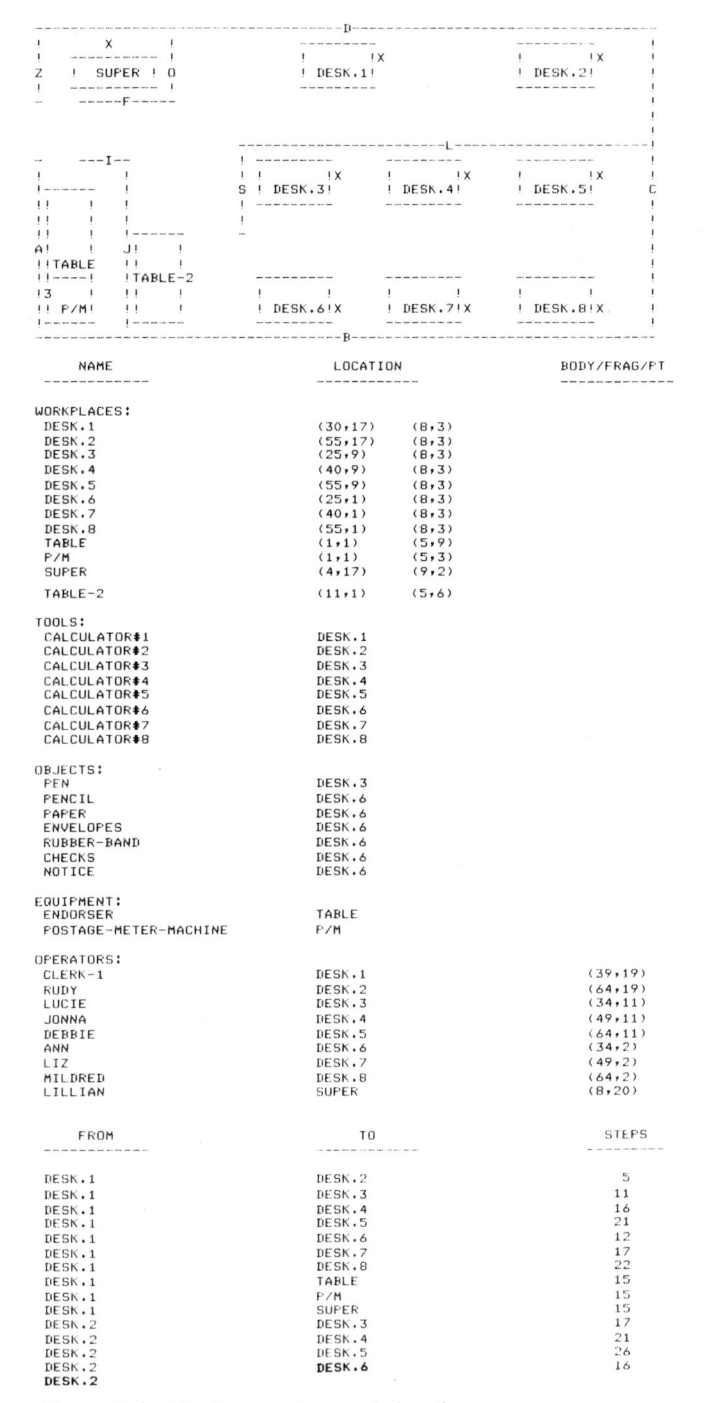

```
----------------------------------D----------------------------------
!        X        !           ---------              ---------       !
!   -----------   !           !       !X             !       !X      !
Z   !  SUPER  !   O           ! DESK.1!              ! DESK.2!       !
!   -----------   !           ---------              ---------       !
-    -----F-----                                                     !
                                                                     !
                                                                     !
                       ------------------------L---------------------!
-      ---I--          ! ---------    ---------    ---------         !
!            !         ! !       !X   !       !X   !       !X        !
!------      !         S ! DESK.3!    ! DESK.4!    ! DESK.5!         C
!!     !     !         ! ---------    ---------    ---------         !
!!     !     !         !                                             !
!!     !     !------   -                                             !
A!     !     J!     !                                                !
!!TABLE      !!     !                                                !
!!----!      !TABLE-2    ---------    ---------    ---------         !
!3     !     !!     !    !       !    !       !    !       !         !
!! P/M!      !!     !    ! DESK.6!X   ! DESK.7!X   ! DESK.8!X        !
!------      !------     ---------    ---------    ---------         !
----------------------------------B----------------------------------
```

NAME	LOCATION		BODY/FRAG/PT
WORKPLACES:			
DESK.1	(30,17)	(8,3)	
DESK.2	(55,17)	(8,3)	
DESK.3	(25,9)	(8,3)	
DESK.4	(40,9)	(8,3)	
DESK.5	(55,9)	(8,3)	
DESK.6	(25,1)	(8,3)	
DESK.7	(40,1)	(8,3)	
DESK.8	(55,1)	(8,3)	
TABLE	(1,1)	(5,9)	
P/M	(1,1)	(5,3)	
SUPER	(4,17)	(9,2)	
TABLE-2	(11,1)	(5,6)	
TOOLS:			
CALCULATOR#1	DESK.1		
CALCULATOR#2	DESK.2		
CALCULATOR#3	DESK.3		
CALCULATOR#4	DESK.4		
CALCULATOR#5	DESK.5		
CALCULATOR#6	DESK.6		
CALCULATOR#7	DESK.7		
CALCULATOR#8	DESK.8		
OBJECTS:			
PEN	DESK.3		
PENCIL	DESK.6		
PAPER	DESK.6		
ENVELOPES	DESK.6		
RUBBER-BAND	DESK.6		
CHECKS	DESK.6		
NOTICE	DESK.6		
EQUIPMENT:			
ENDORSER	TABLE		
POSTAGE-METER-MACHINE	P/M		
OPERATORS:			
CLERK-1	DESK.1		(39,19)
RUDY	DESK.2		(64,19)
LUCIE	DESK.3		(34,11)
JONNA	DESK.4		(49,11)
DEBBIE	DESK.5		(64,11)
ANN	DESK.6		(34,2)
LIZ	DESK.7		(49,2)
MILDRED	DESK.8		(64,2)
LILLIAN	SUPER		(8,20)

FROM	TO	STEPS
DESK.1	DESK.2	5
DESK.1	DESK.3	11
DESK.1	DESK.4	16
DESK.1	DESK.5	21
DESK.1	DESK.6	12
DESK.1	DESK.7	17
DESK.1	DESK.8	22
DESK.1	TABLE	15
DESK.1	P/M	15
DESK.1	SUPER	15
DESK.2	DESK.3	17
DESK.2	DESK.4	21
DESK.2	DESK.5	26
DESK.2	DESK.6	16
DESK.2		

Figure 6.1 Work area, data and sketch.

METHOD DESCRIPTION

- PLACE CHECK FROM DESK 6 TO TABLE 2
- HANDLE COVER ON XEROX-MACHINE SIMO 1
- PUSH DIAL ON XEROX-MACHINE FOR THE NUMBER OF COPIES
- PUSH BUTTON AT XEROX-MACHINE FOR PROCESS
- HANDLE COVER AT XEROX-MACHINE SIMO 1, 2, 3
- PICKUP CHECK
- MOVE COPY FROM XEROX-MACHINE TO DESK 6 WITH SIT

Figure 6.2

To develop a clerical database using the computer, the analyst begins by inputting the work area data—describing the layout of the area, the distances between workplaces (in steps), the equipment being used, the objects being worked on, and any tools that are used. Figure 6.1 illustrates the output that results from generating a workplace. A sketch is produced along with detailed work area data information.

After the work area data has been input, the analyst observes the operation and "speaks" the method description (Fig. 6.2) into a portable dictation device. A typist takes the tape and keys the method description into the computer. The computer will then calculate the amount of time needed to perform the operation. Figure 6.3 provides an illustration of a completed MOST analysis based on the work area data (Fig. 6.1) and the method description (Fig. 6.2).

The completed MOST calculation is then filed in the database simply by using the title of the MOST analysis (see Chap. 9 for a detailed description of the

Figure 6.3 MOST analysis.

```
        COPY   CHECK ON XEROX-MACHINE AT CASH AND MAIL ROOM
PER COPY                                          OFG: 4   04-Dec-78

  1 PLACE CHECK FROM DESK.6 TO TABLE-2
                              A1   B0   G1   A10 B0   P3   A0          1.00        150.
  2 HANDLE COVER ON XEROX-MACHINE SIMO 1
                             <A1>B0   G1   M6   X0   I0   A0          1.00         70.
  3 PUSH DIAL ON XEROX-MACHINE FOR THE NUMBER OF COPIES
                              A1   B0   G1   M1   X0   I0   A0          1.00         30.
  4 PUSH BUTTON AT XEROX-MACHINE FOR PROCESS
                              A1   B0   G1   M1   X10  I0   A0          1.00        130.
  5 HANDLE COVER AT XEROX-MACHINE SIMO 1 2 3
                             <A1  B0   G1  >M6   X0   I0   A0          1.00         60.
  6 PICKUP CHECK
                              A1   B0   G1   A1   B0   P0   A0          1.00         30.
  7 MOVE COPY FROM XEROX-MACHINE TO DESK.6 WITH SIT
                              A1   B0   G1   A10 B10 P1   A0          1.00        230.

                                                    TOTAL TMU                 700.
```

```
CLERICAL TITLE SHEET

        TITLE SHEET ORGANIZATION LIST
        -----------------------------

        ASSEMBLE/DISASSEMBLE
        --------------------

 25- ASSEMBLE  CHECKS ON NOTICE WITH STAPLER AT DESK.6

        EXAMINE
        -------

  8- RESEARCH PAYMENT ON MICROFICHE ON DESK
        NONE

  9- RESEARCH PAYMENT AT HOUSE CARD FILE IN FILE CABINET
        NONE

 10- SORT AND REVIEW PAYMENTS FOR RESEARCH ON MICROFICHE WITH VIEWER
        NONE

 11- CHECK ACCOUNT STATUS FOR ACCOUNT TRANSFER
        NONE

 15- INSPECT CARD FOR COMPLETENESS BEFORE PROCESSING
        NONE

        MOVE
        ----

  3- RETURN PAYMENTS TO MAIL SECTION FOR PROCESSING
        NONE

 22- TRANSPORT MAIL AT CASH AND MAIL ROOM

 23- TRANSPORT CHECKS TO ENDORSER AT CASH AND MAIL ROOM

 29- TRANSPORT  MAIL AT CASH/MAIL ROOM

        OPERATE
        -------

  2- OPEN ENVELOPE ON TABLE WITH AUTOMATIC LETTER OPENER AT MAIL ROOM
        NONE

  4- BUNDLE ENVELOPES FOR DISTRIBUTION WITH HANDS AT MAIL ROOM
        NONE

  6- SORT MAIL IN SLOTS WITH HANDS AT MAIL ROOM
        NONE

 16- BUNDLE FORMS AT CASH AND MAIL ROOM

 18- REMOVE FORMS FROM ENVELOPES AT CASHS AND MAIL ROOM

 19- COPY NUMBERS ON FORM WITH PEN AT CASH AND MAIL ROOM

 20- COPY 5 NUMBERS ON FORM WITH PENCIL AT CASH AND MAIL ROOM

 21- COPY 11 DIGIT NUMBER ON FORM WITH PENCIL AT CASH AND MAIL ROOM

 24- COPY NUMBERS ON FORM WITH PEN AT CASH AND MAIL ROOM

 26- CALCULATE  CHECKS AT DESK.6

 27- COPY  NUMBERS ON FORM WITH PEN AT CASH/MAIL ROOM

 30- SEQUENCE  CHECKS FOR XEROXING AT DESK

 33- COPY  CHECK ON XEROX-MACHINE AT CASH AND MAIL ROOM

        PREPARE
        -------

  1- PREPARE DUPLICATE BILL FOR PAYMENT ON CUSTOMER ACCOUNT
        NONE

  7- MAKE READY TO PREPARE DUPLICATE BILL FOR PAYMENTS RECEIVED WITH NO COUP
     ONS
        NONE

 12- PREPARE WORK-TABLE FOR PROCESSING MAIL
        NONE

 17- PREPARE DESK AT CASH AND MAIL ROOM

 28- MAKE READY  NOTICES FOR STAPLING AT DESK.6

        SURFACE TREAT
        -------------

  5- CLEAN UP AFTER SLITTING ENVELOPES
        NONE

 13- WIFE OFF VIEWER-SCREEN SMUDGES BEFORE RESEARCHING CUSTOMER ACCOUNT
        NONE

 14- CLEAN EXCESS GLUE FROM ENVELOPE BEFORE MAILING
        NONE
```

Figure 6.4 Clerical title sheet.

computerized data filing system). A search can then be made of the database. MOST calculations are organized into a list of all possible operations and suboperations, which could occur for a cost center. This list (a title sheet, Fig. 6.4) is the key to the standards setting process. From this title sheet the analyst can select the various operations that occur and set a standard for a particular job. The output, the final time standard, is presented in three formats: the standard calculation sheet (Fig. 6.5), the time calculation sheet, and the method instruction sheet for the operator (Fig. 6.6). Any suboperation can be traced from its work area layout through a final time standard. In addition to establishing standards, the Clerical Computer System will provide the analyst with performance reports and personnel tables.

The performance report generation program calculates effectiveness, earned hours, coverage, and performance of an individual employee, section, and/or division. The user inputs the number of the standard, the actual quantity of work, the actual time it took to perform the work, and any delay or administrative time. The computer accesses the standard database for the appropriate operational description and the standard time. The program output is a summary of the performance of an employee or section; it contains performance figures, standard hours, nonstandard hours, earned hours, delays, and so on. A weekly report for each section is then stored for the generation of monthly, quarterly, and annual reports.

The staffing requirement program determines the staffing for a cost center. The analyst inputs such variables as the name of the cost center, the number of percentage factions or loaned employees, the planned quantity of work, etc. The program will retrieve data pertaining to the cost center from data files such as time standards and performance reports. The output will include a summary of the standard hours, nonstandard hours, miscellaneous hours, and the projected staff level for the cost center.

Figure 6.5 Standard calculation sheet.

TYPE OF WORK	ELEMENTAL TIME	PERCENT ALLOWANCE	ALLOWANCE TIME	STANDARD TIME
EXTERNAL MANUAL	18910.	15.	2837.	19194.
INTERNAL	(0.)			
PROCESS TIME	0.	0.	0.	0.
STANDARD (TMU/CYCLE)	18910.		2837.	19194.
UNITS PER CYCLE	1			
STANDARD (HOURS/UNIT)				0.192
UNITS PER HOUR @ 100%				5.21

REASON FOR CHANGE -- CHANGE IN PROCEDURES

STEP	METHOD INSTRUCTION		FREQ
1	OPEN ENVELOPE ON TABLE WITH AUTOMATIC LETTER OPE NER	(2)	1
2	SORT MAIL IN SLOTS WITH HANDS	(6)	1
3	TRANSPORT MAIL	(29)	1
	* MAIL TRANSPORT WITH CART		
4	REMOVE FORMS FROM ENVELOPES	(18)	1
5	COPY NUMBERS ON FORM WITH PEN	(27)	1
	* OLD FORM 111		
6	CALCULATE CHECKS	(26)	1
7	SEQUENCE CHECKS FOR XEROXING	(30)	1
8	COPY CHECK ON XEROX-MACHINE	(33)	1
9	TRANSPORT CHECKS	(23)	1

Figure 6.6 Method instruction sheet.

MOST Clerical Systems provides managers with a higher-level, easily learned and understood clerical work measurement tool, along with a meaningful and statistically sound application system. With such tools, that ever-growing portion of our workforce can have realistic goals placed before it. With such goals, performance can be determined and proper assignment of personnel can be accomplished. To improve productivity in clerical organizations is to improve profit. The ease of data analysis and manipulation using the computer has now made the measurement of clerical operations a manageable task.

7
Application of MOST Work Measurement Technique

MOST for Methods Improvement

Preliminary to the actual MOST analysis, the analyst should study the operation with the objective of establishing the most effective method of accomplishing the task. While the "best" method will not always be apparent, every job should be approached with the attitude that "any method can be improved."

The starting point for a study is the information-gathering phase. All important facts concerning the job, such as the workplace layout, tools and equipment, materials, and shop conditions, should be collected and studied in detail. All data should be clearly documented and made easily accessible for future reference. This activity alone should point out many improvement possibilities.

In terms of parameter index values, MOST sequence models give a quantitative description of distances, types of placing activities, tool use frequencies, and so on. During the course of filling out sequences, these index values can serve as indicators for evaluating potential improvements or comparing different methods. All indexes above 3 for A, B, G, and P parameters should be investigated for possible method improvements. For the Tool Use Sequence, index values should reflect the optimum time value based on the choice of tool.

Standard Form for MOST Calculations

Analysis with MOST is simplified by the use of standard calculation forms (see Fig. 7.1). The standard MOST calculation form contains the following six main sections:

1. Identification
2. Area
3. Activity/conditions
4. Method description
5. Basic sequence model analysis section
6. Total time computation

All information necessary to identify, describe, and calculate the standard time for an operation or a suboperation is included on the MOST calculation form.

At the top of the form [see section (1) of Fig. 7.1], space is provided for the 10-digit data bank code (for additional manual coding information, see Chap. 8), the date, the analyst's name, and the number of pages in the analysis. Section (2) of Figure 7.1 is used for indicating the area in which the suboperation analyzed below is found. This could be a general geographical plant area (such as building or department) or working area (such as final assembly, fabrication or turret lathes). Section (3) of Figure 7.1 is a very important section on the form; it is headed activity. The words inserted here determine the size and scope of the suboperation data unit analyzed below. These same words, when manually coded, would appear as the 10-digit code under which the data unit will be filed and retrieved from the data bank. In the MOST Computer Systems, these words alone act as the vehicles by which a suboperation data unit is filed and retrieved from the data bank (see Chap. 9 for more details). It is very important that the words placed in the activity section of the MOST calculation form accurately describe the contents of the suboperation and be uniform and complete. It is suggested that even though suboperations may be manually coded and filed, the title (activity) sentence format for the computer filing system be used; this ensures that a consistent and thorough description of the suboperation is provided. The conditions section of the heading allows the analyst to record any additional descriptive data concerning the suboperation and will aid others in properly identifying it. Special conditions might be:

- For Model X231 only
- Operator must wear special clothing
- For parts up to and including 5 lb.

The left side of the form [Section (4) of Fig. 7.1] is used to record the method description of the activity in a plain-language, logical and chronological sequence. This description can be as detailed or as brief as required. The amount of information placed in the method description section is usually a function of its eventual use; i.e., the description can be used for detailed operator instructions or for an outline of the manual work for time computation only. It is important to note that each method step has *only one* corresponding sequence

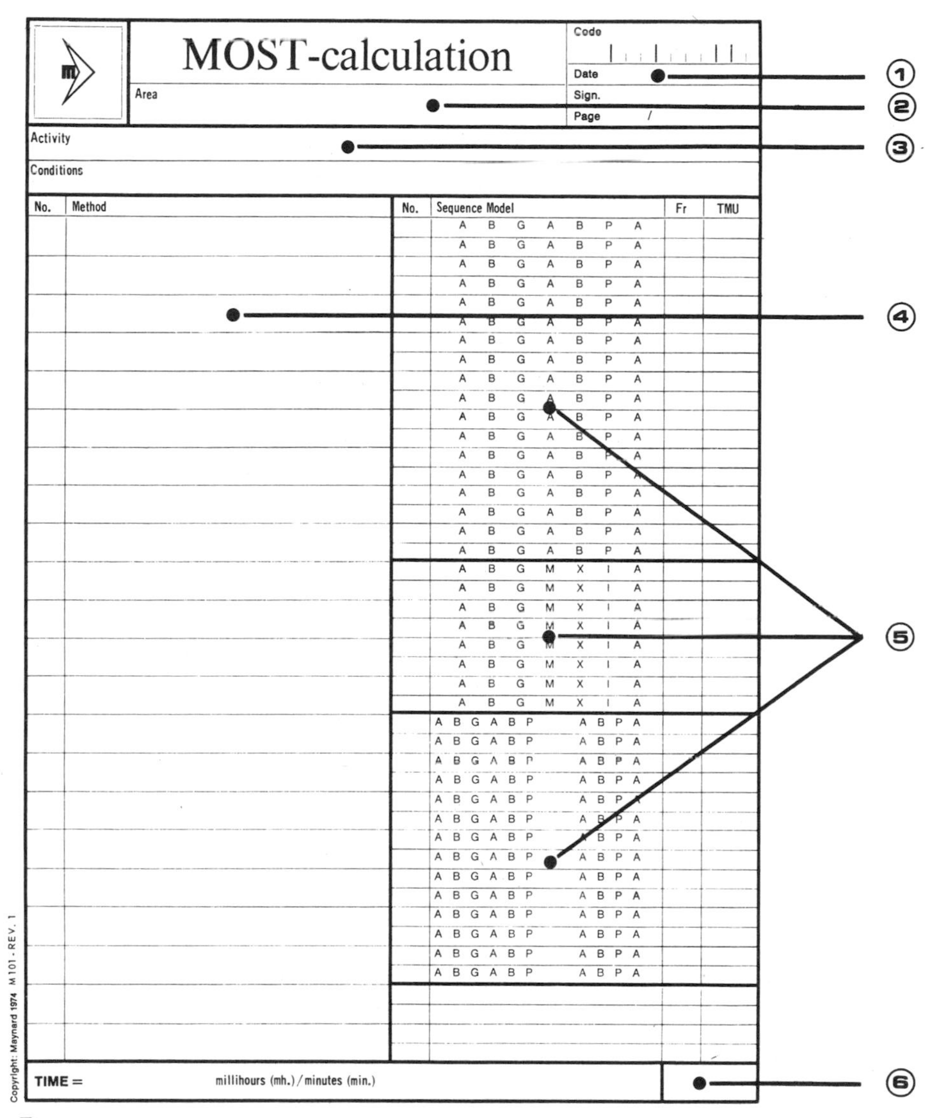

MOST-calculation

Code

Date

Sign.

Page /

Area

Activity

Conditions

No.	Method	No.	Sequence Model	Fr	TMU
			A B G A B P A		
			A B G A B P A		
			A B G A B P A		
			A B G A B P A		
			A B G A B P A		
			A B G A B P A		
			A B G A B P A		
			A B G A B P A		
			A B G A B P A		
			A B G A B P A		
			A B G A B P A		
			A B G A B P A		
			A B G A B P A		
			A B G A B P A		
			A B G A B P A		
			A B G A B P A		
			A B G A B P A		
			A B G A B P A		
			A B G M X I A		
			A B G M X I A		
			A B G M X I A		
			A B G M X I A		
			A B G M X I A		
			A B G M X I A		
			A B G M X I A		
			A B G M X I A		
			A B G A B P A B P A		
			A B G A B P A B P A		
			A B G A B P A B P A		
			A B G A B P A B P A		
			A B G A B P A B P A		
			A B G A B P A B P A		
			A B G A B P A B P A		
			A B G A B P A B P A		
			A B G A B P A B P A		
			A B G A B P A B P A		
			A B G A B P A B P A		
			A B G A B P A B P A		
			A B G A B P A B P A		
			A B G A B P A B P A		

TIME = millihours (mh.)/minutes (min.)

Figure 7.1 MOST calculation form.

model [Section (5) of Fig. 7.1]. Therefore, the method description should be phrased in terms of moving an object(s) or using a tool(s).

The form also contains provisions for applying frequencies to individual sequence models and for the total time calculation for each sequence model. The time for an activity (operation or suboperation) is calculated by simply adding the totals for all the individual sequence models. The grand total is placed in the bottom right-hand corner of the form [Section (6) of Figure 7.1] ; this total time (in TMU) can be converted to decimal minutes, hours, or millihours, and so on.

Note: The total time value reflects the "normal time" for the activity *without any allowances.* Therefore, the time value on the MOST calculation form will have to be multiplied by the appropriate allowance factor (PRD) in order to produce a complete time standard. Also, below the section containing the preprinted sequence models [Section (5) of Fig. 7.1] are four blank lines. These lines are provided for additional sequence models if required. Should one of the preprinted sections above be filled, or should any of the Equipment Handling Sequence Models be required, the additional sequence model(s) would be placed in these four lines. A sampling of completed MOST calculation forms can be found in Appendix B.

Analyst Consistency

Since each parameter or variable pertaining to the three basic sequence models is shown on the calculation form, the analyst will not easily omit or "forget" motions. Each parameter must be filled in or indexed. This forces the analyst to make a decision about which index number to assign. Even nonoccurring motions (index number 0) will require a decision. For this reason, the analyst error of omitting motions will for all practical purposes be eliminated. The result will be a high level of consistency in the application of MOST Systems.

Summary of the MOST Calculation Procedure

The MOST calculation sheet will be filled out as follows.

1. Indicate at the top of the form: code number (from data bank coding system); area of work; activity.
2. Document the method to be analyzed by dividing it into a number of successive steps corresponding to the "natural" breakdown of the activity. Number each step in chronological order.
3. Select one appropriate sequence model for each method step.
4. Indicate the correct index value for each parameter within each sequence model.
5. Add parameter index values together, multiply by 10 (use 100 if applicable), and insert the result in right-hand column to arrive at the sequence time in TMU.
6. For the total activity time in TMU, add all sequence times together and insert the result in the bottom right-hand corner. If desired, these times may be converted to hours, minutes, seconds, or millihours in the bottom left-hand corner of the sheet.

Practical Analysis Procedures

Ideally, observation of two cycles in slow motion will be sufficient to make a MOST analysis. If conditions permit, the operator should first perform the activity from start to finish, allowing the analyst to document the method description. On the next slow-motion cycle, the analyst selects the appropriate sequence model(s) for the corresponding method steps and places index values on each parameter. This procedure requires that the analyst be fully trained, have experience with MOST application, and be thoroughly familiar with the operation.

This approach is, of course, not always possible or even practical. Quite often such calculations have to be made well in advance of the performance of the actual operation in the shop. However, if the method is well established and the analyst possesses complete knowledge of the operation and conditions, the MOST calculation can be made from the office. This requires the use of workplace layouts which include the location and distances of tools, equipment, and parts. The completed analysis should be checked, if possible, by observing the actual operation along with the completed MOST analysis sheet. This procedure is particularly useful for cost estimations of new components and products.

Another analysis procedure that is becoming more common is based on the video taping of operations. Since the MOST Work Measurement Technique is a simple system and a fine measurement method which does not require collection and specification of extremely detailed information, the MOST analysis can often be made directly from observing the operation on a TV monitor. However, the quality of the video tape has to be above average, which will require some practice in the filming of operations.

General Rules for Using MOST

Each sequence model is fixed; no letter may be added or omitted, except as indicated in the sequence model for tool or equipment use.

Index values are fixed; no parameter may carry any index other than 0, 1, 3, 6, 10, 16, 24, 32, 42, 54, etc. For example, there is *no* index number 2.

Each parameter variant must be supported by backup analysis. No index for any parameter may be used unless this backup exists. All elements in the basic MOST system presented in this book are backed up by MTM-1 or MTM-2 analyses.

Updating the MOST Calculation

When evaluating alternative methods or updating existing analyses due to corrections, methods improvements, or the adaptation of these data units to other company divisions or plant situations, it is not necessary to make a completely new analysis each time. Variations from the documented method can be noted on a copy of the original MOST analysis from the data bank simply by changing index values, inserting additional method steps or eliminating method steps. The new method can then be rewritten or typed on a clean calculation sheet and inserted in the data bank.

To illustrate the updating procedure, the following clerical activity will be used. An operator, seated at a desk, stands, picks up letter, and walks 13 steps to a photocopy machine. The cover is raised and the original placed on the glass. The cover is closed. The operator then sets the dial to make one copy. The start button is depressed, and a copying process time of 6 seconds follows. During the process time, the operator gains control of the cover and when the ready light appears, lifts the cover. The original is removed, the cover lowered, and the operator picks up the copy, returns 13 steps to the desk, places the original and the copy on the desk, and sits down.

Figure 7.2 provides the "original" analysis for the above operation. An analyst in another plant observes the method of a similar copying activity and retrieves the original analysis (Figure 7.2) from the company's central data bank. A quick review of the original analysis (Fig. 7.2) tells the analyst that the method for the operation he or she is interested in analyzing differs from the original.

The analyst would then make a copy of Figure 7.2 and replace the original in the data bank. The copy of the original analysis is used as a starting point for updating the calculation to fit his or her particular circumstances. Figure 7.3 illustrates the updating process.

Figure 7.3 reflects the following methods changes.

- The operator's desk is only six steps from the photocopy machine (steps 1 and 10).
- Two dials are manipulated so that 12 copies can be made (step 5).
- The process time increased to 9 seconds (step 6).
- A method step is added to the analysis (step 9′).
- A new total time is generated.
- A new title is applied.

After making all the corrections on the copy of the original analysis, the analyst would then make a smooth copy of the updated analysis (Figure 7.4) and place that in the data bank behind the original analysis. The updating of a MOST calculation would then be completed.

The ease with which MOST calculations can be updated and/or new methods determined is one of the greatest assets of the MOST Work Measurement Technique. It makes simulation and comparison easy and the concept of a data bank of sharable data units a reality.

MOST-calculation

Code	8 0 9 0 0 1 4 2 0 1
Date	10/15/78
Sign.	WMY
Page	1 / 1

Area: CLERICAL

Activity: COPY ONE ORIGINAL AT PHOTOCOPY MACHINE

Conditions: ONE COPY PRODUCED MACHINE MODEL NO. 12345

No.	Method
1	PICK-UP ORIGINAL AND MOVE TO MACHINE
2	OPEN COVER AT MACHINE
3	PLACE ORIGINAL ON GLASS
4	CLOSE COVER
5	SET DIAL FOR 1 COPY
6	PUSH BUTTON TO START COPYING - PROCESS TIME 6 SEC.
7	OPEN COVER
8	REMOVE ORIGINAL
9	CLOSE COVER
10	PICK-UP COPY, RETURN TO DESK AND PLACE ORIGINAL AND COPY ASIDE

No.	Sequence Model	Fr	TMU
1	A_1 B_{10} G_1 A_{24} B_0 P_0 A_0		360
3	A_0 B_0 G_0 A_1 B_0 P_3 A_0		30
8	A_1 B_0 G_1 A_1 B_0 P_0 A_0		30
10	A_1 B_0 G_1 A_{24} B_{10} P_1 A_0		370
	A B G A B P A		
	A B G A B P A		
	A B G A B P A		
	A B G A B P A		
	A B G A B P A		
	A B G A B P A		
	A B G A B P A		
	A B G A B P A		
	A B G A B P A		
	A B G A B P A		
	A B G A B P A		
	A B G A B P A		
	A B G A B P A		
	A B G A B P A		
2	A_1 B_0 G_1 M_3 X_0 I_0 A_0		40
4	A_0 B_0 G_0 M_3 X_0 I_0 A_0		30
5	A_1 B_0 G_1 M_1 X_0 I_0 A_0		30
6	A_1 B_0 G_1 M_1 X_{16} I_0 A_0		190
7	A_1 B_0 G_1 M_3 X_0 I_0 A_0		30
9	A_0 B_0 G_0 M_3 X_0 I_0 A_0		30
	A B G M X I A		
	A B G M X I A		
	A B G A B P A B P A		
	A B G A B P A B P A		
	A B G A B P A B P A		
	A B G A B P A B P A		
	A B G A B P A B P A		
	A B G A B P A B P A		
	A B G A B P A B P A		
	A B G A B P A B P A		
	A B G A B P A B P A		
	A B G A B P A B P A		
	A B G A B P A B P A		
	A B G A B P A B P A		
	A B G A B P A B P A		
	A B G A B P A B P A		

TIME = .68 minutes (min.) — 1140

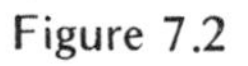

Figure 7.2

MOST-calculation

Code	8,0,9 | 0,0,1,4 | 2 | 0 ~~1~~ 2
Date	10/15/78
Sign.	WMY
Page	1 / 1

Area: CLERICAL

Activity: COPY ONE ORIGINAL AT PHOTOCOPY MACHINE

Conditions: ~~ONE COPY~~ TWELVE COPIES PRODUCED MACHINE MODEL NO. 12345

No.	Method
1	PICK-UP ORIGINAL AND MOVE TO MACHINE
2	OPEN COVER AT MACHINE
3	PLACE ORIGINAL ON GLASS
4	CLOSE COVER
5	SET ~~DIAL~~ DIALS FOR ~~1 COPY~~ 12 COPIES
6	PUSH BUTTON TO START COPYING - PROCESS TIME ~~6~~ 9 SEC.
7	OPEN COVER
8	REMOVE ORIGINAL
9	CLOSE COVER
10	PICK-UP ~~COPY~~ COPIES, RETURN TO DESK AND PLACE ORIGINAL AND ~~COPY~~ COPIES ASIDE
9'	SET DIALS BACK TO ZERO

No.	Sequence Model	Fr	TMU
1	A_1 B_{10} G_1 A_{24} (10) B_0 P_0 A_0		~~360~~ [illegible]
3	A_0 B_0 G_0 (circled A_1) B_0 P_3 A_0		30
8	A_1 B_0 G_1 A_1 B_0 P_0 A_0		30
10	A_1 B_0 G_1 A_{24} (10) B_{10} P_1 A_0		~~270~~ [illegible]
	A B G A B P A		
	A B G A B P A		
	A B G A B P A		
	A B G A B P A		
	A B G A B P A		
	A B G A B P A		
	A B G A B P A		
	A B G A B P A		
	A B G A B P A		
	A B G A B P A		
	A B G A B P A		
	A B G A B P A		
	A B G A B P A		
	A B G A B P A		
2	(circled A_1) B_0 G_1 M_3 X_0 I_0 A_0		40
4	A_0 B_0 G_0 M_3 X_0 I_0 A_0		30
5	A_1 B_0 G_1 M_1 X_0 I_0 A_0	2	~~30~~ 60
6	A_1 B_0 G_1 M_1 X_{16} (24) I_0 A_0		~~180~~ 270
7	A_1 B_0 G_1 M_3 X_0 I_0 A_0		30
9	A_0 B_0 G_0 M_3 X_0 I_0 A_0		30
9'	A_1 B_0 G_1 M_1 X_0 I_0 A_0	2	60
	A B G M X I A		
	A B G A B P A B P A		
	A B G A B P A B P A		
	A B G A B P A B P A		
	A B G A B P A B P A		
	A B G A B P A B P A		
	A B G A B P A B P A		
	A B G A B P A B P A		
	A B G A B P A B P A		
	A B G A B P A B P A		
	A B G A B P A B P A		
	A B G A B P A B P A		
	A B G A B P A B P A		
	A B G A B P A B P A		
	A B G A B P A B P A		

TIME = ~~.76~~ ~~.69~~ .62 ~~hours~~ ~~min.~~ minutes (min.) — ~~1260~~ ~~1150~~ 1030

Figure 7.3

MOST-calculation

Code	8,0,9 | 0,0,1,4 | 2 | 0,2
Date	11/20/78
Sign.	AB
Page	1 / 1
Area	CLERICAL
Activity	COPY ONE ORIGINAL AT PHOTOCOPY MACHINE
Conditions	TWELVE COPIES PRODUCED, MACHINE MODEL NO. 12345

No.	Method
1	PICK-UP ORIGINAL AND MOVE TO MACHINE
2	OPEN COVER AT MACHINE
3	PLACE ORIGINAL ON GLASS
4	CLOSE COVER
5	SET DIALS FOR 12 COPIES
6	PUSH BUTTON TO START COPYING-PROCESS TIME 9 SEC.
7	OPEN COVER
8	REMOVE ORIGINAL
9	CLOSE COVER
10	SET DIALS BACK TO ZERO
11	PICK-UP COPIES, RETURN TO DESK AND PLACE ORIGINAL AND COPIES ASIDE

No.	Sequence Model							Fr	TMU
1	A_1	B_{10}	G_1	A_{10}	B_0	P_0	A_0		220
3	A_0	B_0	G_0	A_1 (circled)	B_0	P_3	A_0		30
8	A_1	B_0	G_1	A_1	B_0	P_0	A_0		30
11	A_1	B_0	G_1	A_{10}	B_{10}	P_1	A_0		230
	A	B	G	A	B	P	A		
	A	B	G	A	B	P	A		
	A	B	G	A	B	P	A		
	A	B	G	A	B	P	A		
	A	B	G	A	B	P	A		
	A	B	G	A	B	P	A		
	A	B	G	A	B	P	A		
	A	B	G	A	B	P	A		
	A	B	G	A	B	P	A		
	A	B	G	A	B	P	A		
	A	B	G	A	B	P	A		
	A	B	G	A	B	P	A		
	A	B	G	A	B	P	A		
	A	B	G	A	B	P	A		
2	A_1 (circled)	B_0	G_1	M_3	X_0	I_0	A_0		40
4	A_0	B_0	G_0	M_3	X_0	I_0	A_0		30
5	A_1	B_0	G_1	M_1	X_0	I_0	A_0	2	60
6	A_1	B_0	G_1	M_1	X_{24}	I_0	A_0		270
7	A_1 (circled)	B_0 (circled)	G_1 (circled)	M_3	X_0	I_0	A_0		30
9	A_0	B_0	G_0	M_3	X_0	I_0	A_0		30
10	A_1	B_0	G_1	M_1	X_0	I_0	A_0	2	60
	A	B	G	M	X	I	A		

No.	Sequence Model	Fr	TMU
	A B G A B P A B P A		
	A B G A B P A B P A		
	A B G A B P A B P A		
	A B G A B P A B P A		
	A B G A B P A B P A		
	A B G A B P A B P A		
	A B G A B P A B P A		
	A B G A B P A B P A		
	A B G A B P A B P A		
	A B G A B P A B P A		
	A B G A B P A B P A		
	A B G A B P A B P A		
	A B G A B P A B P A		
	A B G A B P A B P A		

TIME = .62 minutes (min.) — **1030**

Figure 7.4

Method Levels and Simultaneous Motions

Method level refers to the degree of coordination between the right and left hands during two-handed work. We say that a high method level exists when a large percentage of manual and body motions are performed simultaneously. Obviously it is desirable to have as much work as possible performed at high method levels because of the reduction in time for accomplishing a given amount of work.

The method level at which an activity is performed is determined by its occurrence frequency, i.e., the practice opportunity available to the operator. The more often the activity occurs, the greater the operator's opportunity to improve the method level. If the activity is seldom performed, the short learning period prevents any development of simultaneous skills. For example, with mass production and large batch size operations which allow ample training and practice opportunity, one would expect to find operators using a high percentage of simultaneous motions. On the other hand, job shop and setup activities will most likely be performed with few simultaneous motions. Therefore, method level depends to a large extent on the type of work being performed. Three different method levels are defined for the application of MOST.

1. *High method level* includes all possible simultaneous motions with the right and left hands. The analysis and time for the limiting (longest) hand is allowed. If the analysis for the other hand is shown, the time value must be circled indicating that this value is not included in the total. The following activity occurs simultaneously with and is "limited" by another activity.

A_1 B_0 G_1 A_1 B_0 P_1 A_0 (40) (Time 0 TMU)

In this case, the sequence time in the right column is circled to indicate it is not included in the total.

2. *Low method level* involves no simultaneous motions. The analysis and time for both hands must be allowed.

A_1 B_0 G_1 A_1 B_0 P_1 A_0 40 TMU

3. *Intermediate method level* refers to a method performed partially with simultaneous motions. For example, the action distance "within reach" to two objects may be performed simultaneously with both hands, but gaining control and placing two objects simultaneously may not be possible. In the MOST analysis, the appropriate parameter(s) are circled to indicate they are performed simultaneously and should be excluded from the sequence model time calculation. In the following activity, a portion of the sequence model (the reach to get the object) is performed simultaneously with another activity.

(A_1) B_0 G_1 A_1 B_0 P_1 A_0 30 TMU

In this case, the circled portion of the sequence model is not included in the time calculation as it is "limited" by another activity.

Examples

The activity place two pins in assembly is analyzed using three different method levels. A pin is picked up by each hand and placed in the assembly with adjustments.

1. High method level: both hands work simultaneously.

RH	$A_1\ B_0\ G_1\ A_1\ B_0\ P_3\ A_0$	60
LH	$A_1\ B_0\ G_1\ A_1\ B_0\ P_3\ A_0$	(60)
		60 TOTAL TMU

2. Low method level: both hands work separately.

RH	$A_1\ B_0\ G_1\ A_1\ B_0\ P_3\ A_0$	60
LH	$A_1\ B_0\ G_1\ A_1\ B_0\ P_3\ A_0$	60
		120 TOTAL TMU

3. Intermediate method level: only the Get occurs simultaneously.

RH	$A_1\ B_0\ G_1\ A_1\ B_0\ P_3\ A_0$	60
LH	($A_1\ B_0\ G_1$) $A_1\ B_0\ P_3\ A_0$	40
		100 TOTAL TMU

As can be seen from the example, there is a wide variation in the total time for each method level; therefore, one of the analyst's most important considerations in a work measurement situation is to represent the correct method level in the analysis. This relationship between method and standard (time) should always be emphasized in MOST analysis work and should be based on the theory that the greater the practice opportunity, the higher the method level.

Index Value Development for Special Tools or Situations

One of the most important features of the MOST system is the provision for developing special parameters or index values for activities unique to local conditions.

Special Tools

The Tool Use data cards were designed to provide accurate parameter values for a wide range of common tools found throughout industry. While the majority of tools can be analyzed using the data from Tables 4.1 and 4.2 (either directly or by comparison), there may be special tools used in an operation that are not covered by any of these Tool Use groups. If the tool is infrequently used, the basic sequence models (General and Controlled Move) can, of course, be used to analyze its use; but, if the tool is frequently used, it may be desirable to develop special tool use parameters specifically for the tool.

Three alternatives are available to the analyst for describing the use of those tools *not found* in the tool use tables:

1. Identify the method employed, compare it with existing data, and select an appropriate index value from a similar tool use method. (It is always the method of using a tool, not the name of the tool, that determines the parameter value.)
2. Make a detailed MOST analysis using a combination of General and Controlled Move Sequences.
3. For frequently used tools, develop a special parameter with index values based on an MTM-1, MTM-2, or time study analysis.

Alternative 1: Compare Method and Use Existing Data Frequently, a special tool will resemble another tool in appearance as well as the method employed. A corkscrew, for example, which requires the use of wrist actions, also looks very much like a small T-wrench. Therefore, as this alternative suggests, the activity to "turn" a corkscrew into a cork (e.g., with six wrist actions) can be analyzed using the Fasten/Loosen data for a small T-wrench:

$$A_1\ B_0\ G_1\ A_1\ B_0\ P_3\ F_{16}\ A_0\ B_0\ P_0\ A_0$$

$(1+1+1+3+16) \times 10 = 220$ TMU

Since light pressure is needed to start the corkscrew, a P_3 is required for the tool placement. The removal of the cork is then described using the General Move Sequence. (Notice the G_3 for "disengage.")

$$A_0\ B_0\ G_3\ A_1\ B_0\ P_1\ A_0$$

$(3+1+1) \times 10 = 50$ TMU

(The removal of the cork from the corkscrew will require another Tool Use Sequence Model.)

Alternative 2: Analyze the Method Using General and Controlled Move Sequences If an appropriate index value is unavailable after comparing a special tool method with the existing data, the activity can be analyzed using parameters from General and Controlled Move Sequences. For example, the method of using a crank-operated hand drill does not seem to fit any of the tools listed in Tables 4.1 or 4.2. However, a detailed MOST analysis can be made by breaking the complete drilling activity down into its basic subactivities. The analysis for using a hand drill to make a hole in a wooden block with eight revolutions of the crank handle would require three sequence models. (1) Get and place hand drill to a mark on the block:

$A_1 \ B_0 \ G_1 \ A_1 \ B_0 \ P_3 \ A_0$

(1+1+1+3) X 10 = 60 TMU

(2) Get hold of handle and drill hole with eight cranking actions:

$A_1 \ B_0 \ G_1 \ M_{16} \ X_0 \ I_0 \ A_0$

(1+1+16) X 10 = 180 TMU

(3) Remove and set hand drill aside:

$A_0 \ B_0 \ G_3 \ A_1 \ B_0 \ P_1 \ A_0$

(3+1+1) X 10 = 60 TMU

Note: This alternative should be used for those tools *infrequently found* in use because of the analysis detail involved.

Alternative 3: Develop Index Values for the Tool One of the most useful features of the MOST Work Measurement Technique is the provision for the development of index values for special parameters. This feature is particularly applicable when a *frequently used* tool (or applicable method) is not found in the tool use data. The index value determination procedure first requires that the tool use method be analyzed using either MTM-1, MTM-2, or time study. Index values are then assigned to the tool according to the MOST time interval into which each analysis falls.

Consider, for example, an assembly operation in which a spiral screwdriver is frequently used. The MTM-2 analysis for this activity might be:

Analysis	TMU	Description
GW6	3	Accumulate muscle tension (6 lb.)
PA6	6	Power stroke
PW10	1	With 6 lb. resistance
PA6	6	Return stroke
	16	TMU per tool action

With an additional 14 TMU (MTM-2, apply pressure) for final tightening, the tool use parameter for the spiral screwdriver can be expressed algebraically:

$t = 16N + 14$

where t = time per tool action in TMU and,
N = number of tool actions.

This formula, representing the spiral screwdriver parameter, can be completed for various numbers of tool actions (N) and converted to the proper MOST index values using the time intervals from Appendix A, Table A.

Example: Using the above formula, the time value for one tool action (30 TMU) falls within the interval between 18 and 42 TMU which corresponds to the index number 3. This could be calculated individually for N actions. A less tedious way is to complete the formula for three separate values for N (for example: $N = 1, N = 5, N = 11$) and plot these values on an Index Value Determination Form (Figure 7.5).

Steps to develop index values for the spiral screwdriver example using the Index Value Determination Form are as follows.

1. Perform MTM-1 or MTM-2 analysis.
2. Develop algebraic formula: $t = 16N + 14$.
3. Choose three separate values for N and work the formula.

$t = 16(1) + 14 = 30$ TMU
$t = 16(5) + 14 = 94$ TMU
$t = 16(11) + 14 = 190$ TMU

4. Using an Index Value Determination Form, identify and supply the number of variables to the x axis of the form (Fig. 7.5).
5. Plot the TMU values for the three formulas worked in step 3 above. (Use the TMU or seconds scale on the form; Fig. 7.5.)
6. Connect the points plotted with a straight line. (*Note:* Should the line curve or veer sharply, further detailed analysis should be performed to determine the variable influencing the shape of the line; Fig. 7.5.)
7. Where the plotted line crosses one of the horizontal lines printed on the form (the upper limit for each index range), draw a vertical line. This vertical line divides the number of tool actions into various index ranges (Fig. 7.5). A simple matter of placing these values in tabular form results in the development of a supplementary index value table for a spiral screwdriver (see Table 7.1).
8. Additional values could always be obtained by working the formula and assigning the index number from Table A.1 of Appendix A.

If the spiral screwdriver were used to fasten a screw with ten tool actions, the MOST analyst could now use one Tool Use Sequence Model (for index values have been developed that can be "plugged in" to the sequence model). The analysis would appear:

$A_1 \; B_0 \; G_1 \; A_1 \; B_0 \; P_3 \; F_{16} \; A_1 \; B_0 \; P_1 \; A_0$

$(1 + 1 + 1 + 3 + 16 + 1 + 1) \times 10 = 240$ TMU

The preceding situation dealt with the development of index values for a spiral screwdriver based on a detailed MTM-2 backup analysis. Backup for index values can also be developed using time study. Situations that lend themselves to time study backup analyses are such activities as polishing, grinding, painting, or glueing, or any other activity involving a short process time, i.e., using power tools or office machines. Index values should be developed for these situations when found frequently enough in the operations of a company or plant to justi-

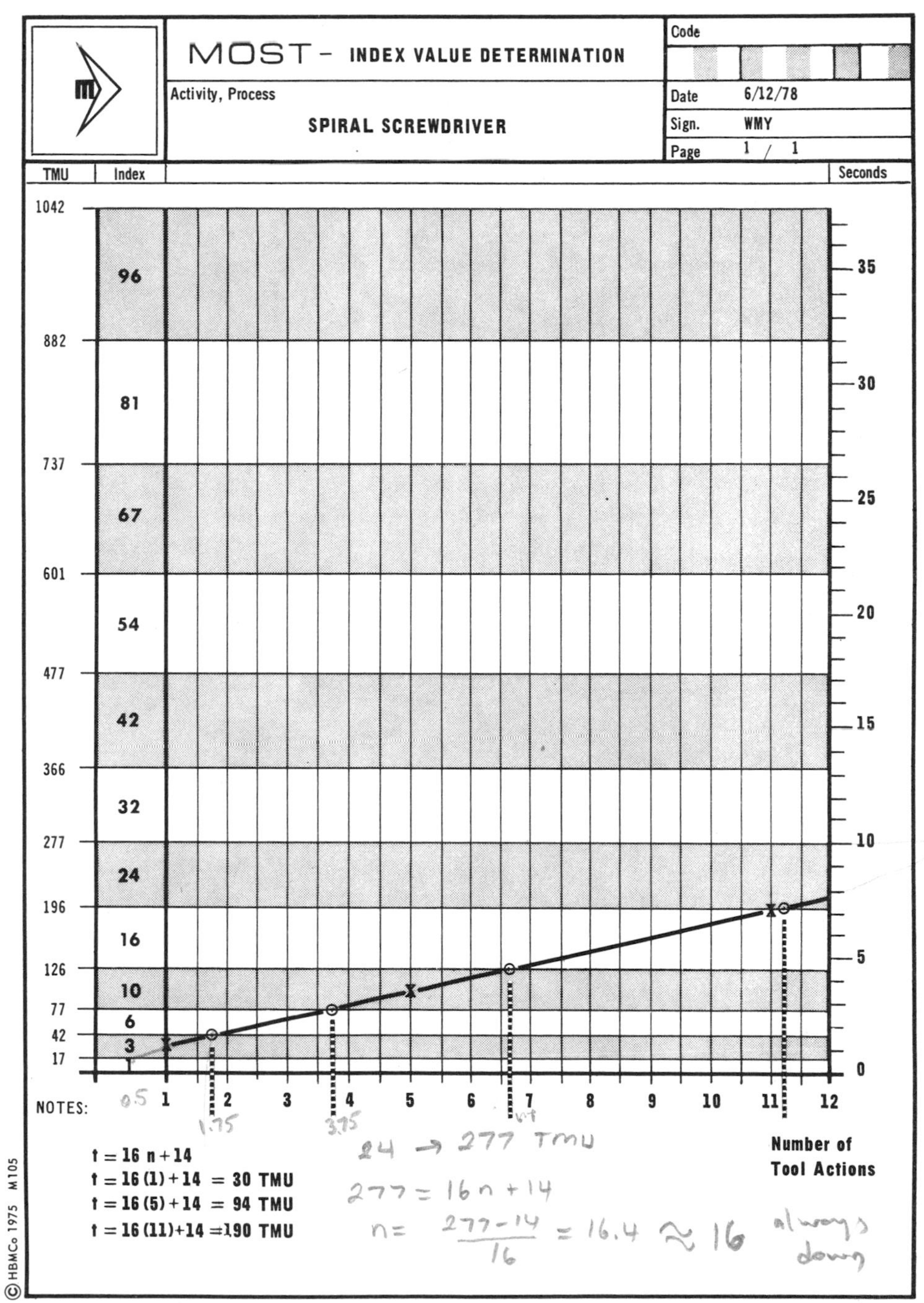
MOST - INDEX VALUE DETERMINATION
Code
Activity, Process
SPIRAL SCREWDRIVER
Date 6/12/78
Sign. WMY
Page 1 / 1
TMU
Index
Seconds
1042 882 737 601 477 366 277 196 126 77 42 17
96 81 67 54 42 32 24 16 10 6 3
35 30 25 20 15 10 5 0
0.5 1 2 3 4 5 6 7 8 9 10 11 12
1.75
3.75
Number of Tool Actions
NOTES:
t = 16 n + 14
t = 16 (1) + 14 = 30 TMU
t = 16 (5) + 14 = 94 TMU
t = 16 (11)+14 = 190 TMU
24 → 277 TMU
277 = 16n + 14
n = (277 − 14)/16 = 16.4 ≈ 16 always down
©HBMCo 1975 M105

Figure 7.5

Table 7.1 Spiral Screwdriver*

INDEX NUMBER	NUMBER OF TOOL ACTIONS
1	-
3	1
6	3
10	6
16	11

*Read "up to and including."

fy the time taken to develop such values and when consistency of application is required.

To determine index values using time study, the unit of the variable should be specified, the proper time studies performed, and the results plotted on the Index Value Determination Form. For example, the times for polishing might be in seconds per square foot; this would be plotted, and a supplementary data table for polishing per square foot would be developed. To use the data, values from this table could then be applied to the Tool Use Sequence Model and placed under the Surface Treat (S) parameter.

MOST Analysis Decision Diagram

To aid the MOST analyst, a MOST Analysis decision diagram is provided as Figure 7.6. This diagram will lead the analyst through all the basic thought processes and decisions that need to be considered in order to arrive at a thorough and consistently applied MOST analysis. In the diagram, the boxes indicate a process or operation and the diamonds indicate that a binary decision is required. Simply follow the process and your decisions through the diagram to properly complete an analysis of a suboperation.

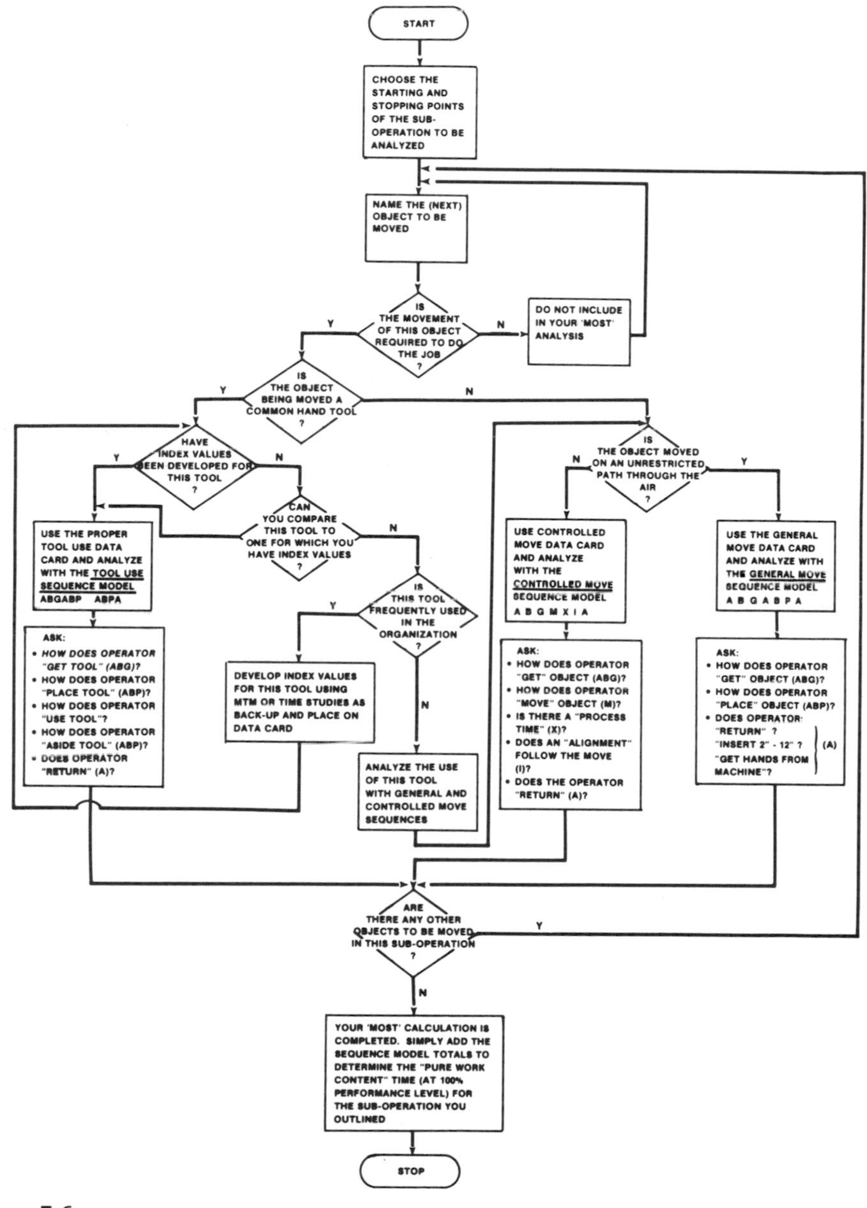
START
CHOOSE THE STARTING AND STOPPING POINTS OF THE SUB-OPERATION TO BE ANALYZED
NAME THE (NEXT) OBJECT TO BE MOVED
IS THE MOVEMENT OF THIS OBJECT REQUIRED TO DO THE JOB ?
DO NOT INCLUDE IN YOUR 'MOST' ANALYSIS
IS THE OBJECT BEING MOVED A COMMON HAND TOOL ?
HAVE INDEX VALUES BEEN DEVELOPED FOR THIS TOOL ?
CAN YOU COMPARE THIS TOOL TO ONE FOR WHICH YOU HAVE INDEX VALUES ?
USE THE PROPER TOOL USE DATA CARD AND ANALYZE WITH THE TOOL USE SEQUENCE MODEL ABGABP ABPA
ASK:
• HOW DOES OPERATOR "GET TOOL" (ABG)?
• HOW DOES OPERATOR "PLACE TOOL" (ABP)?
• HOW DOES OPERATOR "USE TOOL"?
• HOW DOES OPERATOR "ASIDE TOOL" (ABP)?
• DOES OPERATOR "RETURN" (A)?
IS THIS TOOL FREQUENTLY USED IN THE ORGANIZATION ?
DEVELOP INDEX VALUES FOR THIS TOOL USING MTM OR TIME STUDIES AS BACK-UP AND PLACE ON DATA CARD
ANALYZE THE USE OF THIS TOOL WITH GENERAL AND CONTROLLED MOVE SEQUENCES
IS THE OBJECT MOVED ON AN UNRESTRICTED PATH THROUGH THE AIR ?
USE CONTROLLED MOVE DATA CARD AND ANALYZE WITH THE CONTROLLED MOVE SEQUENCE MODEL A B G M X I A
ASK:
• HOW DOES OPERATOR "GET" OBJECT (ABG)?
• HOW DOES OPERATOR "MOVE" OBJECT (M)?
• IS THERE A "PROCESS TIME" (X)?
• DOES AN "ALIGNMENT" FOLLOW THE MOVE (I)?
• DOES THE OPERATOR "RETURN" (A)?
USE THE GENERAL MOVE DATA CARD AND ANALYZE WITH THE GENERAL MOVE SEQUENCE MODEL A B G A B P A
ASK:
• HOW DOES OPERATOR "GET" OBJECT (ABG)?
• HOW DOES OPERATOR "PLACE" OBJECT (ABP)?
• DOES OPERATOR: "RETURN" ? "INSERT 2" - 12" ? "GET HANDS FROM MACHINE"? (A)
ARE THERE ANY OTHER OBJECTS TO BE MOVED IN THIS SUB-OPERATION ?
YOUR 'MOST' CALCULATION IS COMPLETED. SIMPLY ADD THE SEQUENCE MODEL TOTALS TO DETERMINE THE "PURE WORK CONTENT" TIME (AT 100% PERFORMANCE LEVEL) FOR THE SUB-OPERATION YOU OUTLINED
STOP
Y
N

Figure 7.6

8
Systematic Development of Engineered Labor Time Standards

Two Alternatives: Final Time Standards or Suboperation Data

MOST Work Measurement Systems unquestionably provides all the necessary features of an effective work measurement system either for the direct establishment of final time standards for operations or for the development of suboperation data for portions of operations. As is indicated above, MOST provides several capabilities superior to conventional techniques.

There are many situations in which the direct approach of calculating time standards with MOST is both the fastest and most economical route to take. An assembly line with relatively short-cycled operations and few product variations is an almost perfect example of an area within which the time standards can be set without any intermediate steps or data levels. On the other hand, and in the majority of cases, some type of "building blocks" will have to be developed prior to the final stage of calculating time standards. The reason for this complication is primarily economical and to a certain extent practical.

Since the calculation of final time standards quite often is a continuous process, it is obviously important to minimize the costs for this activity. Therefore, the procedures used must be quick and simple to apply. However, an adequate accuracy in the allowed time values must be retained. Even though MOST has a combined accuracy and speed characteristic superior to any other known system, the speed of MOST is still far from sufficient to calculate every time standard economically on a direct basis.

It is, therefore, quite necessary to introduce one or more levels between the basic work measurement system (MOST) and the final time standard. Traditionally, units or elements developed on levels below final time standards have been designated "standard data." Standard data units could range all the way from very short time values of a few TMU to elements several minutes long. All standard data elements were grouped and classified at various data levels. With MTM-1 as a basis, for example, it has been common to define three to five different data levels.

The build up of standard data was always made "bottom up," meaning that the smallest elements were developed first, with subsequent combining of such elements into larger and larger standard data units. Logically and technically this approach was sound; economically and practically it was very costly and time consuming.

With the introduction and application of the MOST Work Measurement Technique this standard data development procedure has been completely reversed. With MOST one always starts at the operation level (equal to final time standards) and then continues from the top down, if necessary, by breaking operations into suboperations. By doing that, only those data that are actually required to set the standard have to be developed. Normally, each completed MOST calculation form represents one complete operation or suboperation. It is for this reason that the activity title assigned to the MOST calculation is so important–it identifies the data unit. It seldom happens when using MOST, but it is possible that more than one level below the operation level will be needed. Such situations are frequent for data developments in industries like shipbuilding and heavy engineering where the value of a final time standard may exceed 100 hours. In such cases, two or three levels of data may be required.

The general rule (for other than direct analyses) is that all MOST calculations produced be called *suboperations.* All suboperations are classified as *suboperation data* and placed on a level immediately below the operation level. This is also true in cases where "higher levels" of suboperations are common; these higher levels are called *combined suboperations.* The concept of standard data is, therefore, not applicable for MOST Systems, and neither are the words standard data being used in association with MOST.

There is no need to develop elemental standard data units with MOST; each sequence model is an element in itself. Usually, one MOST analysis (suboperation) consists of 5 to 15 method steps, or sequence models, corresponding to an identical number of individual elements in terms of standard data. Certainly, variations from this average may occur, but only in rare cases does a MOST analysis consist of only one sequence model or reflect the analysis of one element.

The distribution of sequence models per MOST calculation is shown in Figure 8.1 for a sample of manufacturing industries. MOST is so fast in application that the development, filing, retrieval, and possible revision of small elements (standard data units) has proven to be an inefficient and cumbersome procedure to follow. Because of the minute detail of MTM-1, the standard data

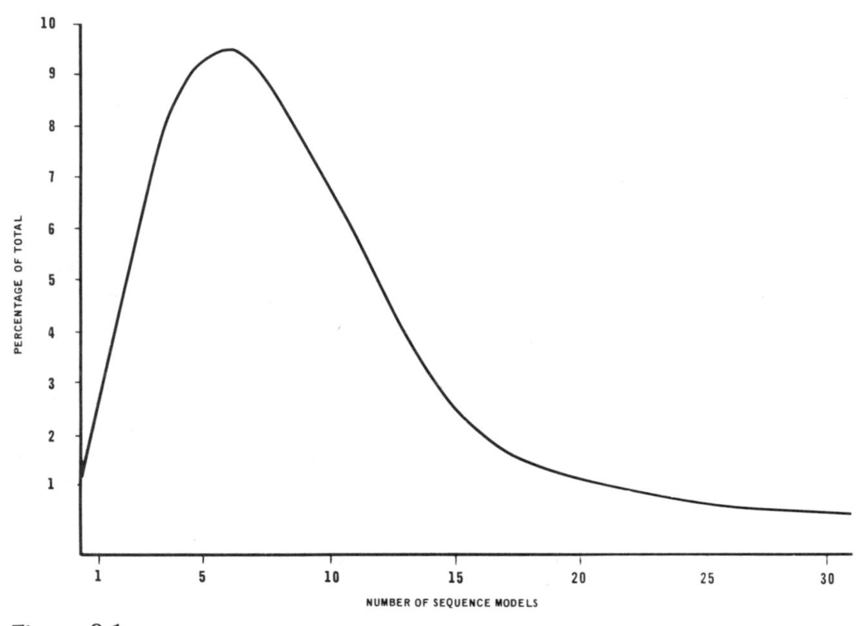

Figure 8.1

building block approach was the only possible solution in the application of that system. This principle was also true for the higher-level systems such as MTM-2, MTM-3, GPD, and USD, but it no longer holds true when using the MOST systems.

The direct approach with MOST is a quicker and simpler way to reach the final time standards. The key is the ease with which complete suboperations can be analyzed or existing ones updated to fit particular situations when using MOST.

Filing and Retrieval of Suboperation Data

It is obvious that the suboperation data units being developed from MOST analyses have to be organized in some way. For several reasons, it must be easy to retrieve one specific unit from a large quantity of suboperations. A unit might be needed and reviewed for the purpose of reference, method (and time standard) change, application in a different area, method evaluation, and improvement or comparison. Therefore, systematic filing of suboperation data units in a data bank using a uniform and simple coding system has become an integral part of an effective standard time system. Each unit can be assigned a unique code or number consisting of, for instance, 10 digits. These digits, divided into three groups or categories, can give a brief description of the content of the suboperation.

- Activity (three digits)
- Object/component/equipment/tool (four digits)
- Method level (one digit)

The last two digits of the 10-digit code may be reserved for a sequential running number. Such a number is necessary to establish a unique code for each unit. To make the coding simple and practical, only 10 digits may be used. These digits are usually not enough to describe all the variations of a suboperation that may occur. On the other hand, a coding system with 20 to 25 digits is very cumbersome and impractical to use.

The following examples will illustrate the application of a 10-digit code number for suboperation data units.

1. Set up and tear down a machine vise.
   Code: 103 0310 4 01

Activity: Set up and tear down	103
Object group: Holding devices	03
Object (within object group): Vise	10
Method level (infrequent activity)	4
Sequential running number	01
(first analysis made having the indicated eight digits)	

2. Assemble resistor on PC-board.
   Code: 101 5522 2 02

Activity: Assemble	101
Object group: Electronic components	55
Object: Resistor	22
Method level (frequent activity)	2
Sequential running number	02
(Second analysis with this eight-digit code number)	

3. Read instruction sheet for billing procedures (clerical activity)
   Code: 703 0011 3 05

Activity: Read	703
Object group: General	00
Object (within object group): Instruction sheet	11
Method level (infrequent activity)	4
Sequential running number	05

The suboperations can be filed according to activity and object group or object. Using this procedure, there will be a cross register, and the search for a unit can be made with regard to either activity or object.

All MOST analyses (suboperation data units) are arranged sequentially in the data bank for easy retrieval. The search for data is a completely manual process even if the word *data bank* implies that a computer is involved. There are, however, many advantages with a computerized data bank, as is indicated in Chapter 9. It must be emphasized, though, that a well-organized and well-maintained manual data bank can be an effective and practical instrument in the development and maintenance of time standards.

Approaches to the Calculation of Final Time Standards

The establishment of time standards is quite often a continuous process, requiring a certain amount of labor. Time standards must be set for the operations pertaining to new and changed products and components as well as to any change in the work conditions. Some companies may have to calculate and maintain hundreds of thousands of time standards, since (1) the total number of operations is very large; (2) product additions and variations are common; and (3) new facilities, equipment, and tools are constantly being added.

Depending on the quantity of new time standards and the effectiveness of the applied procedures for setting them, the required applicator capacity can be estimated. In order to keep the cost-benefit ratio for this function on an acceptable level, it is important that the procedures and data application formats for the establishment of time standards be properly developed and refined.

There are many ways of setting time standards, even if we limit ourselves to so-called engineered time standards. An engineered time standard is based on documented work conditions and a method specification, in addition to the time value. On the other hand, an estimated time standard consists of a time value only. No backup is usually available for an estimated time standard.

The most common data application formats for engineered time standards can be classified in three major groups:

1. *Standard time tables* indicating fixed time values and final time standards for the operations in an area.
2. *Work sheets* that normally contain all possible activities or suboperations within an area. The final time standard will be calculated by "checking off" actual suboperations for a specific operation and adding the time values for the suboperations together. Frequently, process time and allowances will also have to be included (Fig. 8.2).
3. *Spread sheets* are made up of benchmark operations slotted into preset time groups. The time standard will be established by comparing the work content of the actual operation with any one of the benchmarks on the spread

WORK SHEET - EXAMPLE

m>	Area: ASSEMBLY	Code
		Date: 75.04.12
	Operation: OIL PUMP PXA-4810	Sign. A.A.

SUBOPERATIONS														Total MH Set Up	Total MH Per Piece
Preparation															
Simple	101 1 100					Normal 102 300				Complicated	103		600	100	
Transport to work place															
Elevate											104	1	77	77	
Crane											105		87		
		Pos. No.					Frequence				No	Fr	Mh		
Transport	To work place	1				Elevate	1				201	1	77		77
						Crane					202		87		
Handling	Turn work piece	5				Crane	1				203	1	49		49
	Work piece up/down	2				Crane	1				204	1	59		59
Assemble	Screw and/or nut					≤ 1/4"					205		10		
		3	4			> 1/4"	4	2			206	6	14		84
	Subassembly										207		42		
	Stud										208		12		
	Pin	6					2				209	2	8		16
	Locking wire										210		15		
	Locking pin/washer	4					2				211	2	12		24
	With torque					≤ 30 kpn					212		7		
						> 30 kpn					213		14		
Cut						≤ 10 holes Diam ≤ 5/8"					214		2		
Chamfer	Hole					> 10 holes Diam ≤ 5/8"					215		1		
						≤ 10 holes Diam > 5/8"					216		4		
						> 10 holes Diam > 5/8"					217		2		
											218				
	Radius										219		40		
											220				
Mark	Diameter					Diam. ≤ 8"					221		6		
						Diam. > 8" ≤ 16"					222		8		
						Diam. > 16"					223		10		
	With scriber										224		2		
											225				
TOTAL														177	309
OPERATION TIME INCLUDE ALLOWANCES 10%														195	340

Figure 8.2 Example work sheet.

SET-UP .195 Hours
OPERATION .340 Hours

sheet. The mean value of the time group into which the actual operation falls is the final time standard that will be allowed (Fig. 8.3).

While *standard time tables* are used in situations where the number of both final time standards and method variations is low, *spread sheets* are, on the other hand, being applied in areas where the total number of final time standards is

Craft: Automotive	Code	1420.07
	Signature	VH
Task Area: Electrical System	Date	6/23
	Page	1/7

Group: A	B	C	D
Time: 0 – (0.10) – 0.150	0.150 – (0.20) – 0.250	0.250 – (0.40) – 0.500	0.500 – (0.70) – 0.900
ENGINE TIMING SET W/LIGHT 1420.07 - 07 .1350	BATTERY - R/R FORD CAB OVER OR EQUIV. 1420.07 - 44 .1960	SWITCH - TOGGLE - R/R 1420.07 - 48 .2785	BEARING - ALT. - R/R 1420.07 - 71 .7813
SEAL BEAM (ONE) 1420.07 - 10 .0808	BOLT - BATTERY CLAMP R/R (ONE) UNDER HOOD 1420.07 - 45 .1791	TIMING & DWELL - CHECK DELCO - 6 CYL. MOST EQUIPMENT 1420.07 - 56 .2982	
LIGHTS & INSTRUMENTS - CHECK 1420.07 - 12 .0534	CHARGE BATTERY (FAST) HOOK UP & REMOVE 1420.07 - 50 .1507	DIST. W/GOU. R/R ALL TRUCKS 1420.07 - 57 .3053	
POINTS - ADJUST. DELCO WINDOW TYPE 1420.07 - 13 .0811	TAIL LITE ASSY. - R/R ALL EQUIPMENT 1420.07 - 51 .1574	S/PLUGS - R/R - MOST - 6 CLY. 1420.07 - 58 .3253	
CLEARANCE LITE ASSY. R/R 1420.07 - 14 .0619	POINTS & COND. - R/R & ADJUST. 1420.07 - 55 .2135	ELECTRIC PLUG - R/R TRAILER CABLE ALL 1420.07 - 64 .3682	
CLEARANCE LITE ASSY. R/R USE LADDER 1420.07 - 14A .0957	SWITCH - TURN SIGNAL - R/R UNIVERSAL TYPE 1420.07 - 65 .1561	SWITCH - TURN SIGNAL - R/R FORD 1420.07 - 69 .3800	

Figure 8.3 Example spread sheet.

large and the method variations are difficult to control, for instance in such areas as maintenance and warehousing.

A final time standard can thus be obtained directly from a time table or a spread sheet by either a selection or a comparison. Establishing a final time standard from a *work sheet* is always a matter of selecting the relevant suboperations, assigning an order of sequence and applicable frequencies, and computing the time value. Usually, specifications of the method, tooling, and process data are being produced at the same time, mainly for instructional purposes. Both *standard time tables* and *spread sheets* are obviously very fast to apply once they have been properly developed. Therefore, the output rate of time standards is high, and a relatively limited applicator capacity will be needed.

Work sheets can be developed to any degree of detail desired. A detailed work sheet requires more capacity to calculate the time standards. Therefore, it is important to pay a great deal of attention to how work sheets are being designed. The suboperations on the work sheet should be arranged and grouped so that it will be easy for the applicator to locate them, as well as to make quick decisions as to which suboperations to select.

By using a statistical approach in determining allowed deviations or tolerances for each suboperation, the number of suboperations listed on the work sheet can frequently be reduced through combinations of two or more of them. In the process of designing a work sheet, it is rather common that the total number of suboperations may be reduced by as much as two-thirds. Such simplication of the work sheet will minimize the selection and decision time during the calculation procedure and make the applicator more efficient, thus cutting the cost per standard.

There are additional benefits. It is, for example, quite possible to control the output accuracy of the time standards produced, as well as to improve the consistency in the output by applying a familiar statistical formula for standard deviations theory (for further explanations, see Appendix A).

Documentation of Work Conditions

It has been pointed out several times throughout this book that engineered time standards have to be well documented with regard to methods and work conditions. This principle is as valid for MOST Systems as it is for any other predetermined motion time system. Without a proper documentation of actual backup data, the time standard can be considered neither engineered nor defendable should a dispute ever arise. It is also important that the documentation supporting a time standard reflect realistic and actual work conditions. Since work conditions change continuously, updating and maintenance of the time standards to match the prevailing conditions at any point in time become critical issues throughout the installation and application phase of the standards development process.

The documentation of work conditions is similar to taking a photograph of the shop or office where the work is being performed. The work conditions we talk about here can be subdivided into the following categories.

- Facilities: machines, equipment, and tools
- Work flow: such as material flow, work area, and workplace layouts

- Technical and manual methods
- Standard practices

These and other work conditions can preferably be described and documented for each work area in a manual with a standardized format. Such a manual will not only be regarded as a reference for backup material for the time standards, but it will also serve as a handbook for the supervisor or foreperson. All necessary information concerning the conditions within the supervisor's domain will be readily available, the objective being to help the supervisor better perform his or her job of work management.

Consistently developed, these manuals of work conditions or Work Management Manuals, can find many more applications. They can be used in training situations for industrial engineers, supervisors, and workers. They can be used as precedents for the development of new manuals in similar areas, and they can be used as reference material in the event of union disputes over time standards.

Systematic Development of Time Standards

The subject of systematic development of time standards is by itself large and complicated enough to fill a book. Only a few basic concepts and principles, together with the important features of the MOST Application System, have been outlined in this section. It has been necessary to exclude from this text most of the detailed procedures that make up a well-designed standard time system.

MOST Work Measurement Systems is like the powerful engine that is being built into an automobile or truck. The engine is needed to make the vehicle move and operate, but the steering, the wheels, and other components actually make the automobile or truck do what you expect it to do: transport people or material. Likewise, it is the components and features of a standard time system or, in this case, MOST Application Systems that really will help you get "the most out of MOST." The standard time system developed by H. B. Maynard and Company is structured to produce engineered labor time standards with MOST Work Measurement Systems as the engine or drive train and the MOST Application Systems as the steering, wheels, cab, etc.

The systematic application procedures for MOST consist of a set of rules under which MOST should be applied, as well as a documentation package for improving the work management and thus the productivity of the industrial engineer and the company. In any case, such application procedures must be developed or acquired, provided they do not already exist in the company, to produce engineered labor time standards.

9
MOST Computer Systems

Introduction

Throughout the past decade the use of the computer in industry has spread rapidly. Scores of computer terminals are located throughout plants to enhance the flow of information within or between various departments. In manufacturing engineering, the primary uses of computers have been for process and inventory control and for directing the flow of operational procedures to the factory floor.

Although these applications by manufacturing engineering departments have proven very useful, there are many other benefits still to reap. The computer's speed, accuracy, and ability to sort and collate large amounts of data can be used to relieve the engineer of many routine tasks. With the advent of the minicomputer's on-line capabilities, the engineering department's access to the computer, so often a problem with large-scale company computers, is no longer a barrier.

MOST Computer Systems addresses one area of the industrial engineering realm—the establishment of labor time standards based on the MOST Systems described in this book. In the majority of companies today, most of the work involved in gathering data and preparing time standards is still being done manually by the industrial engineer. Yet, many of these tasks can be performed more quickly and accurately by a computer, thus freeing the engineer to focus on more productive tasks. Although MOST as a manual system is consistent and fast, MOST Computer Systems offers even greater speed and uniformity of application.

A Totally Integrated System

Any properly functioning system is designed such that the various parts are interactive and form a unified whole. MOST Computer Systems is designed in this manner. Emphasis is placed on a "total system" to set and maintain complete labor time standards. The various system program components are linked together to accomplish four basic functions (see Fig. 9.1):

1. Development of data
2. Storage of data
3. Standard calculation
4. Storage of standards

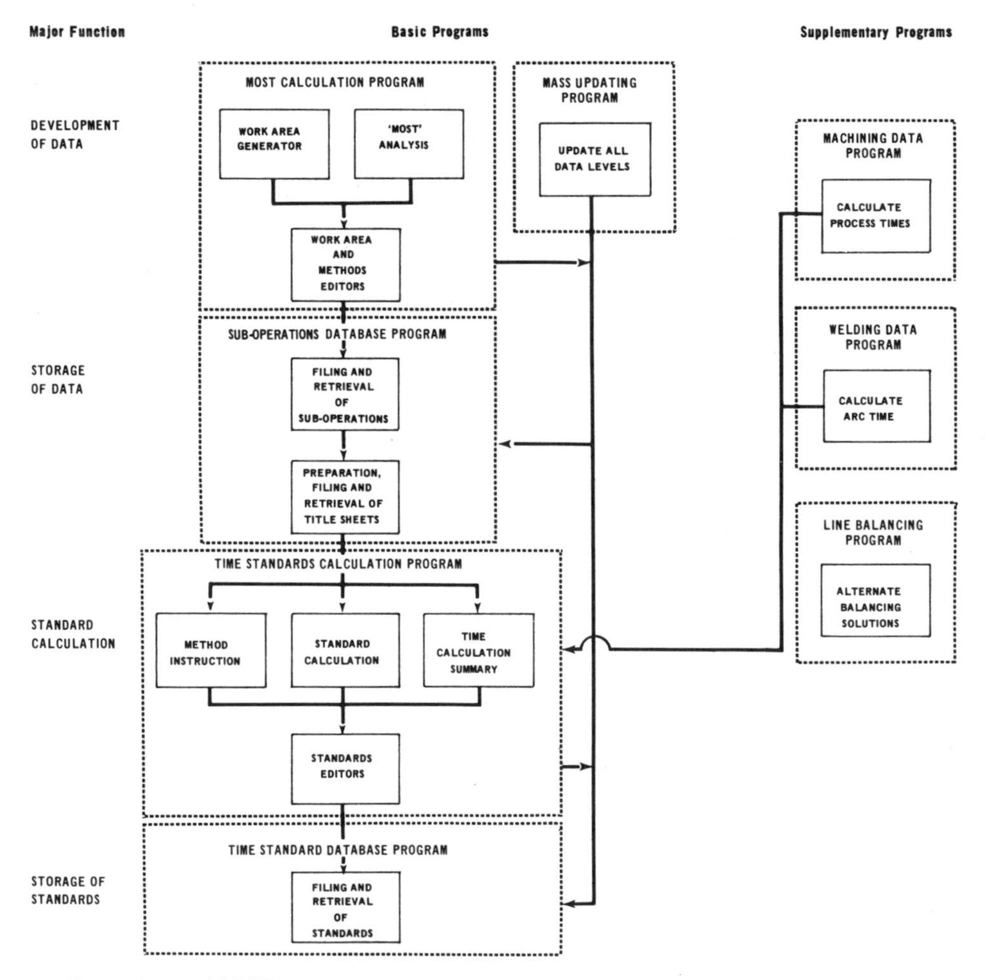

Figure 9.1 MOST computer system, program components.

This linkage allows the engineer the opportunity to follow a unit of data (an operation or suboperation) through the system, all the way from the appropriate workplace layout to the time standard(s) of which it is a part. The system therefore provides a complete audit trail along with the proper documentation to produce an "engineered labor time standard." This design is the basis for the mass updating program which enables the industrial engineer to keep "all standards" current at the introduction of *any* change, i.e., workplace, method, or process. This feature provides an organization the opportunity to say, "Our standards are current and accurate!" no matter what or how extensive the change. To accomplish the four basic functions listed above, MOST Computer Systems is made up of a number of individual programs linked together and illustrated in Figure 9.1. They are the:

- MOST Calculation Program, which includes: The Work Area Generator, The Methods Analysis routine, The Time Calculation (MOST analysis) routine, Editing routines for the workplace and method description, and, printing routines
- Suboperation Database Program, which includes: Filing and retrieval routines for MOST analyses and the corresponding work areas; Retrieval of suboperation data by multiple combinations of categories; Preparation, filing, and retrieval of title sheets.
- Time Standards Calculation Program, which includes: selection and grouping of operations and/or suboperations from the title sheet; the Standards Calculation routine; preparation of the Operator Instruction sheet; the allowancing of personal, rest, and delay times; the Entry of Process times; the Standard Editing routines
- Time Standards Database Program, which includes: the Filing and retrieval of final time standards by multiple combination of categories; the linkage to the suboperation database
- Mass Updating Program, which includes updating and maintenance routines for both databases

Even though all these components are necessary for a total system for the development and maintenance of time standards, they are produced as modules. Therefore, some can be grouped together or used independently to fill particular requirements. For instance, the MOST Calculation Program and its subroutines can be run as a separate unit.

In addition to the program components above, there are several supplementary programs that interact with the basic programs and provide process times where applicable. They are the:

- Machining Data Program, which includes: calculation of machine process times; the selection of machine feeds and speeds and the recommended tool grades

- Welding Data Program, which includes calculation of the arc time for automatic, semiautomatic, or manual welding operations
- Line Balancing Program, which generates alternate line-balancing solutions.

And, for those situations that dictate the use of other analytical tools, Regression Analysis and Multi-Man/Machine Analysis Programs are available.

A Few New Terms

While it is based entirely on the MOST Work Measurement System and the MOST Application System (Chap. 8), the use of the computer has forced the development of a few new terms which, defined in the accompanying table, will appear throughout this chapter.

Term	Definition
CRT	Cathode ray tube, a video display terminal
Data	Any characters entered into the computer to be processed; data maintained in the computer on files is called stored data
Disk	A piece of equipment that magnetically records and stores data
Disk Storage	Storage of data on a disk
Hardware	The tangible, physical aspects of a computer system; the equipment components that make up the computer system
Input	Information entered into the computer; the activity to enter data
Key word	Special preprogrammed words, used in a method description, that are functional in the computer's development of sequence models
Minicomputer	A device capable of solving problems by accepting input data, processing that data, and

Term	Definition
	producing output; a multicomponent computer, noted for quick response time; generally smaller and less expensive than large "main frame" computers
On-line	The status of a communication link between a remote terminal (CRT) and the computer, i.e., when "on-line," direct communication can be accomplished
Operator instruction sheet	Provides a listing of the Method Instructions for performing an operation
Output	Data, which are usually formatted, obtained from a computer
Program	Computer software that acts on input and produces output; a set of instructions
Software	A set of written instructions for processing data and/or solving problems
Standards calculation sheet	A summary listing of all manual, process and allowance times for a particular operation
Terminal	Hardware with which a user supplies input to and receives output from the computer
Time calculation summary	A listing of the times for individual suboperations with appropriate frequencies; a summary of the mathematical standard time calculation
Time Sharing	The computer's ability to interact with many users running many programs during the same time span
Title Sheet	A listing of the titles of MOST calculations used to set standards

Development of Data

The Work Area Generator and MOST Calculation Program

The first step in calculating any engineered labor time standard is the documentation of the workplace. Many so-called engineered labor time standards are established without proper workplace documentation. In those cases when a question regarding a standard time arises, insufficient information concerning the work area leads to a situation where the industrial engineer cannot defend a calculated time. When broken down to the basic input elements, any calculated time reflects only two factors, the *work area* and the *method* performed. MOST Computer Systems is based on these two inputs. Therefore, MOST Computer Systems provides complete documentation of the work area, including a sketch (see Fig. 9.2), and a comprehensive, logical, and easily understood method description.

The Workplace

To perform a MOST analysis using the computer, an industrial engineer gathers the work area information and "speaks" that information into a small tape recorder or dictating device. The tape is handed to a typist, who keys the information into a cathode ray tube (CRT) terminal. The work area is now complete, and a printout of the work area and the work area data can be easily obtained. Figure 9.2 is an illustration of such a printout for a typical multispindle vertical drill. Important work area information input to the computer is: workplace names and tools and their locations; objects and their locations; equipment and its location, along with a process time, if appropriate; operator(s) and their starting location(s); body motions always associated with particular work places and the distance (in steps) between work places.

The Method

After describing the work area, the industrial engineer improves and specifies the method to perform the particular operation or suboperation. The method is spoken into the tape recorder in a plain-language sentence format. Figure 9.3 is an illustration of a method description for placing a part in a fixture of the multispindle vertical drill. Each method step begins with a "key word" that designates to the computer the sequence model to use, as well as the values for the Gain Control (G) and Place (P) parameters. For example, the key word Place means General Move, G_1, P_3. The word Move means General Move, G_1, P_1. These words, as well as the entire format for the method description, totally conform to familiar English sentence structure and industrial engineering terminology and are therefore very easy to learn. Given these key words and the previously entered distances, body motions, and locations (A and B values), the computer can handily calculate an engineered time standard.

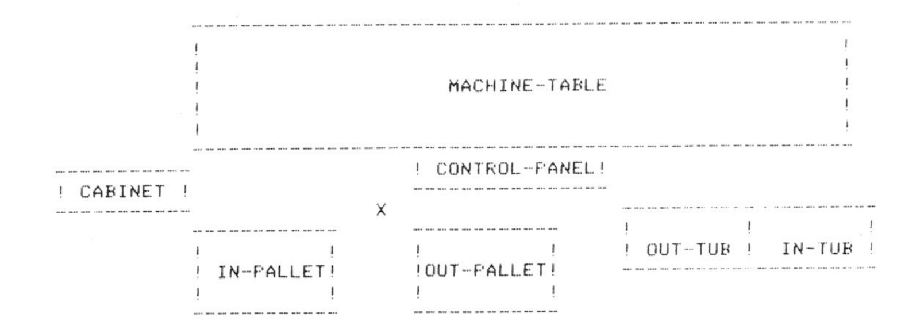

Name	Location		Body/Frag/PT
WORKPLACES:			
MACHINE-TABLE	(14,13)	(53,6)	
CONTROL-PANEL	(32,11)	(15,2)	
IN-PALLET	(14,5)	(11,4)	
OUT-PALLET	(32,5)	(11,4)	
IN-TUB	(59,7)	(10,3)	
OUT-TUB	(49,7)	(10,3)	
CABINET	(3,10)	(10,2)	
TOOLS:			
AIR-HOSE	MACHINE-TABLE		
BOX-END-WRENCH	MACHINE-TABLE		
BEAR-CLAW	MACHINE-TABLE		
PENCIL-GRINDER	MACHINE-TABLE		
OBJECTS:			
RAW-PART	MACHINE-TABLE		
PART	IN-PALLET		FRAG
FIN-PART	OUT-PALLET		FRAG
MOVE-TICKET	CABINET		
WORK-ORDER-PACKET	CABINET		
EQUIPMENT:			
JIB-CRANE	MACHINE-TABLE		
FIXTURE	MACHINE-TABLE		
CLAMPS	MACHINE-TABLE		
PIN	MACHINE-TABLE		
UNIVERSAL-VISE	MACHINE-TABLE		
LID	MACHINE-TABLE		
BUTTON	CONTROL-PANEL		
OPERATORS:			
OP1	MACHINE-TABLE		(29,10) B

From	To	Steps
MACHINE-TABLE	CONTROL-PANEL	1
MACHINE-TABLE	IN-PALLET	4
MACHINE-TABLE	OUT-PALLET	4
MACHINE-TABLE	IN-TUB	1
MACHINE-TABLE	OUT-TUB	1
MACHINE-TABLE	CABINET	2
CONTROL-PANEL	IN-PALLET	4
CONTROL-PANEL	OUT-PALLET	4
CONTROL-PANEL	IN-TUB	2
CONTROL-PANEL	OUT-TUB	3
CONTROL-PANEL	CABINET	4
IN-PALLET	OUT-PALLET	5
IN-PALLET	CABINET	3
OUT-PALLET	CABINET	7
IN-TUB	OUT-TUB	1
IN-TUB	CABINET	10
OUT-TUB	CABINET	9

Figure 9.2 Workplace data.

METHOD DESCRIPTION INPUT
• PLACE PART FROM IN-TUB TO FIXTURE
• PUSH SLIDE CLAMP AT FIXTURE
• FASTEN 2 NUTS AT FIXTURE WITH 4 ARM-TURNS USING BOX-END WRENCH AND ASIDE.
• FASTEN SCREW FASTENER AT FIXTURE 1 SPIN USING FINGERS

Figure 9.3

The taped method description is transcribed to a CRT by the typist. The computer then calculates the operation or suboperation time (Fig. 9.4). The industrial engineer's main function, in the work measurement process, becomes one of establishing and inputting a proper work area layout and an efficient and workable method. Once these basics have been established, the computer takes over the task of calculating times. This frees the industrial engineer for more productive tasks. An illustration of the entire basic data entry process appears as Figure 9.5

One of the major advantages of the computer system is the consistency offered by the key-word approach to method descriptions. Calculation errors and errors from selecting wrong values from the charts are totally eradicated. Also, by focusing upon the work area and method, the industrial engineer is directly performing the tasks with the most influence on the productivity of the production function—the very reason for the industrial engineering department's existence.

Figure 9.4 MOST analysis.

```
    LOAD  PART IN FIXTURE WITH BOX END WRENCH AT
MULTI SPINDLE VERTICAL DRILL 2000
PER PART
                                                          OPG: 2  28-Dec-78

  1 PLACE PART FROM IN-TUB TO FIXTURE
                                A3  B0  G1  A3  B0  P3  A0          1.00       100.
  2 PUSH SLIDE CLAMP AT FIXTURE
                                A1  B0  G1  M1  X0  I0  A0          1.00        30.
  3 FASTEN 2 NUTS AT FIXTURE WITH 4 ARM-TURNS USING
     BOX-END-WRENCH AND ASIDE
       A1  B0  G1  A0  B0  (P3  A1  F10 )A1  B0  P1  A0  (2)        1.00       320.
  4 FASTEN SCREW FASTENER AT FIXTURE 1 SPIN USING FINGERS
                A1  B0  G1  A1  B0  P1  F1  A0  B0  P0  A0          1.00        50.

                                                      TOTAL TMU                500.
```

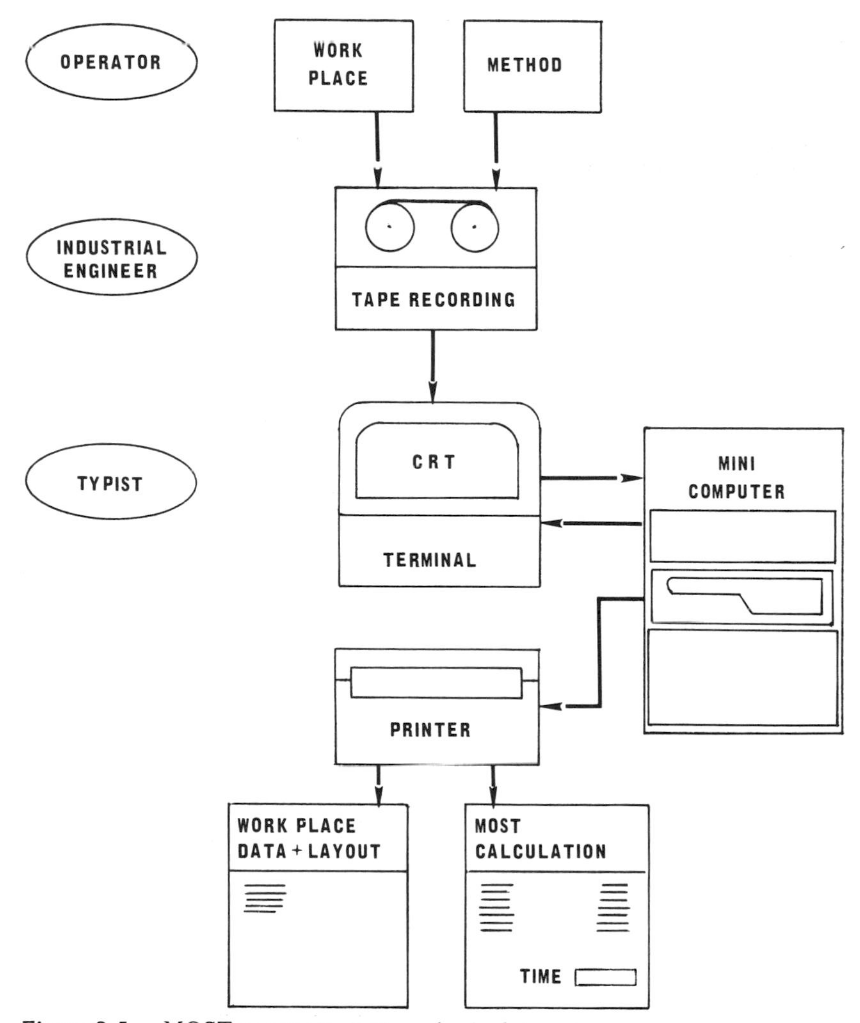

Figure 9.5 MOST computer systems basic data entry.

A key element of maintaining any data system is the ability to edit data both during the input phase and immediately after the calculation has been made, or at any time later. MOST Computer Systems is an on-line system: all input errors are immediately diagnosed and can be corrected instantly. Once the data have been processed, however, the MOST Computer Systems editors provide a useful tool to simulate or effect changes in the work area layout and/or methods. In the work area, for example, locations, distances, and body motions may be changed, workplaces and articles added, and fixed process times adjusted. In the method descriptions, steps may be inserted, deleted, or completely changed. The system then reprocesses the data, based on the changes, and a new suboperation is immediately available. The editors are easy both to learn and to apply. Both the "new" and the "old" analyses are stored in a holding area on disk until the engineer decides which data to transfer to the database. The editors function for both permanently and temporarily stored data. The benefits of the editing feature are:

1. Simulation of changes in either work area or method is an easier task, as all calculations are made automatically. The engineer becomes the interpreter of the results rather than actually performing the new calculations.
2. Transfer of suboperation data from one area in a plant to another or between plants is readily accomplished by simply editing the workplace or methods to meet the conditions of the new application.
3. The engineer can establish prototype workplace layouts and methods and use the editors for adding details. This procedure shortens engineering time spent on analyzing similar situations.
4. Changes are easy to implement, and the impacts of change are instantly apparent.

Storage of Data

Suboperation Database Program—Filing and Retrieval of Suboperation Data

One of the critical components of establishing basic data for use in establishing time standards is the ability to retrieve it at will under any system, manual or computer. Data retrieval is dependent upon the way the data are coded. A prime advantage of a computerized filing and retrieval system is the computer's ability to manipulate and sort vast amounts of data. In manual systems, coding fields are usually kept to a minimum (Chap. 8), and multiple sorts of the data are nearly impossible to attain because of the difficulties encountered in trying to manipulate large amounts of data by hand. These physical constraints simply do not exist for a computer. In fact, a computer can sort words as quickly as it can sort numbers! These characteristics initiated the development of a unique data filing and retrieval system which is tailored to the needs of the industrial engineering department.

Using MOST Computer Systems, all suboperation data units are filed by a plain-language sentence containing the activity, object, product or equipment, tool, work area origin, size or capacity of the workplace, and work area number. Appropriate prepositions placed between the above-listed categories formulate a "title" for the MOST suboperation unit. In the process of affixing a title to the

MAJOR	CATEGORIES	• ACTIVITY(IES)	LOAD
		• OBJECT(S) / COMPONENT(S)	PART
		in / on / for	in
		• EQUIPMENT / PRODUCT	FIXTURE
		with	with
		• TOOL(S)	BOX END WRENCH
		from / to / at	at
		• SIZE / CAPACITY	LARGE
		• ORIGIN	MULTI SPINDLE VERTICAL DRILL
		• MACHINE NUMBER	2000
SUPPLEMENTARY	DATA	• per	per
		• UNIT	PART
		• OCCURRENCE FREQUENCY GROUP	OFG 2
		• DATE (COMPUTER ASSIGNS)	August 16, 1977
		• SPECIAL USER CATEGORY(IES)	(APPLICATOR, PLANT, ETC.)
		• CONDITIONS	(RESTRICTIONS, SPECIAL APPLICATIONS, ETC.)

Figure 9.6 MOST computer systems filing program categories for suboperation data units.

suboperation, the industrial engineer is also determining the categories by which that unit will be filed and subsequently retrieved. This double feature will therefore minimize the necessary input of data for the filing procedures. The input expressed here as a suboperation title will also be used later for other purposes within the system.

After approval of the selected categories by a data coordinator, the suboperation is filed in the database by each of the selected categories. Retrieval is accomplished by selection of any one or any combination of the categories. For example, the data unit described by the title and categories shown in Figure 9.6 may be retrieved by any combination of categories described in Figure 9.7.

Once the desired suboperaton has been located, there is an option of formats available to examine the data, for example:

Title, special conditions, method description, sequence models, and TMU total
Title, special conditions, and method description
Title, special conditions, and TMU total
Work area layout

SEARCH FOR	CATEGORY TYPE	RETRIEVED ANALYSES
ACTIVITY:	LOAD	ALL "LOAD" OPERATIONS IN THE DATABASE INCLUDING THE DESIRED ANALYSIS
ACTIVITY: PRODUCT EQUIPMENT:	LOAD FIXTURE	THE LOADING OF ALL OBJECTS IN FIXTURES FOR ALL MACHINES IN ALL PLANTS
ACTIVITY: PRODUCT / EQUIPMENT: WORK AREA ORIGIN:	LOAD FIXTURE MULTI-SPINDLE VERTICAL DRILL	THE LOADING OF ALL OBJECTS IN FIXTURES AT ALL MULTI-SPINDLE VERTICAL DRILLS
ACTIVITY: OBJECT: PRODUCT / EQUIPMENT: TOOLS: SIZE / CAPACITY: WORK AREA ORIGIN: WORK AREA NUMBER:	LOAD PART FIXTURE BOX-END WRENCH LARGE MULTI-SPINDLE VERTICAL DRILL 2000	ONLY THE LOADING OF PARTS IN A FIXTURE WITH A BOX-END WRENCH - LARGE MULTI-SPINDLE VERTICAL DRILL NO. 2000
WORK AREA ORIGIN:	MULTI-SPINDLE VERTICAL DRILL	ALL SUBOPERATIONS FOR MULTI-SPINDLE VERTICAL DRILLS. THIS SEARCH TECHNIQUE IS UTILIZED FOR DETERMINING TOTAL COVERAGE AT A WORK AREA AND FOR RETRIEVING ITEMS FOR PLACEMENT ON A TITLE SHEET

Figure 9.7

If the engineer locates an analysis that can be used in another operation, plant, or location, the data can be edited and tailored to exactly fit the new situation.

There are numerous advantages with the suboperation filing and retrieval routines available in the MOST Computer Systems.

1. Data units are literally at the industrial engineer's fingertips—no more lost files or data without codes.
2. Work area layouts can be retrieved with every analysis.
3. The flexibility of the search technique allows many combinations of data to be retrieved at one time.
4. Use of the database and editor combination provides the ability to prefabricate an analysis for an entirely new part or operation.
5. Linkages between all types of data units and between the suboperation data and standards databases are an integral part of the system, allowing instant update and crosschecking. A suboperation can always be tracked to a final time standard.
6. Uniformity, the key to a good filing system, is demanded and, in fact, enforced.
7. The consistent filing system creates uniformity among departments and plants, allowing for easier retrieval of data.
8. Plants within the same company can share data and even the same database at the very same time.

Creating a Title Sheet

After analyzing the range of situations concerned with one operation or with varying operations involving one type of machine, the analyst will want to arrange the operations or suboperations into a format that can be used for the actual setting of time standards. Such formats are commonly referred to as worksheets. Since the computer performs all mathematical calculations, there no longer is a need for a worksheet. The MOST Computer Systems has replaced the worksheet with a Title Sheet. The Title Sheet merely lists and groups by title all the suboperations of a particular operation. The Title Sheet is created by searching the database to identify all suboperations and then simply creating a list of those which are appropriate. Figure 9.8 is an example of a Title Sheet for the multispindle vertical drill. The completed Title Sheet is itself filed in the database under the category "work area origin."

Standard Calculation: The Time Standards Calculation Program

The final objective of MOST Computer Systems is to arrive at a complete time standard. This is accomplished by searching the database for the correct

```
MULTI SPINDLE VERTICAL DRILL

        TITLE SHEET ORGANIZATION LIST
        -----------------------------

        MOVE
        ----

  38- LOAD  PART IN MACHINE VISE AT MULTI SPINDLE VERTICAL DRILL

 134- LOAD  PART IN FIXTURE WITH BOX END WRENCH AT
      MULTI SPINDLE VERTICAL DRILL 2000

 137- LOAD  PART IN FIXTURE WITH HOIST AT MULTI SPINDLE VERTICAL DRILL
        FOR PARTS GREATER THAN 80 LBS.

 135- UNLOAD  PART IN MACHINE VISE AT MULTI SPINDLE VERTICAL DRILL

 136- UNLOAD  PART IN FIXTURE WITH BOX END WRENCH AT
      MULTI SPINDLE VERTICAL DRILL

 138- UNLOAD  PART IN FIXTURE WITH HOIST AT MULTI SPINDLE VERTICAL DRILL
        PARTS GREATER THAN 80 LBS.

 108- STACK  PART ON TUB AT MULTI SPINDLE VERTICAL DRILL
        TO CORNER OF TUB

        OPERATE
        -------

  44- START  BUTTON ON CONTROL PANEL AT MULTI SPINDLE VERTICAL DRILL

        PREPARE
        -------

  45- MAKE READY  TAG ON WORKBENCH WITH PENCIL AT
      MULTI SPINDLE VERTICAL DRILL

        SURFACE TREAT
        -------------

  46- CLEAN  PART ON FIXTURE WITH AIR HOSE AT
      MULTI SPINDLE VERTICAL DRILL
        3 SQ. FT.

  47- DEBURR  PART ON TUB WITH GRINDER AT MULTI SPINDLE VERTICAL DRILL
```

Figure 9.8 Multispindle vertical drill title sheet.

Title Sheet and selecting the appropriate suboperations from the Title Sheet, specifying their correct sequence, applying the appropriate frequencies, and indicating whether the suboperation is internal or external to another suboperation or process. The Time Standards program then produces three pieces of output.

1. Method instruction sheet for the operator (Fig. 9.9)
2. Standard calculation sheet (Fig. 9.10)
3. Time calculation sheet (Fig. 9.11)

With careful analysis of Figures 9.2, 9.4, and 9.8, through 9.11 you can follow the suboperation "load part in fixture with box-end wrench at large multispindle vertical drill #2000" through from its work area layout (Fig. 9.2) to the final time standard (Fig. 9.10).

Up to this point, the discussion has centered upon the calculation of manual times. As seen in Chapter 3 on the Controlled Move Sequence Model, short and relatively fixed process times are usually included directly in the MOST analysis itself under the X (process time) parameter. Longer and variable process times must frequently be calculated separately. There are several supplementary programs included in MOST Computer Systems that can be utilized in process time calculation, depending upon company needs: the Machining Data program, the Welding Data program, and special calculations and storage of frequently occurring process times for nonmachining-type operations (foundry, clerical, etc.). Appropriate allowance factors are applied to the manual and process times. The result is the final time standard.

Figure 9.9 Method instruction sheet.

```
PART NO. A-111-222          R.L.  QQ          OPER NO. 000012

OPER. DSCRP. DRILL HOLES IN PLATE

PART NAME FLAT PLATE                    COST CENTER 0000123

ISSUE NO.  1                            MACH. CENTER 001111

DATE    1-29-79                         ASSET NO. 000135

APPLICATOR WMY                          CREW/MACH. 1/1

PLANT 02


STEP  METHOD INSTRUCTION                                                    FREQ

1     LOAD  PART IN FIXTURE WITH BOX END WRENCH       (  134)               1
2     START  BUTTON ON CONTROL PANEL                  (   44)               1
3     DRILL 5 HOLES                                   ( MACH)               1
       * 495 RPM   5.9079 IPM
4     CLEAN  PART ON FIXTURE WITH AIR HOSE            (   46)    (INT) 1
       * INTERNAL TO MACHINE TIME
5     UNLOAD  PART IN FIXTURE WITH BOX END WRENCH     (  136)               1


        TOTAL TMU - MANUAL              990.0
        TOTAL TMU - PROCESS             806.0
        STANDARD HOURS PER PIECE          0.023
        PIECES PER HOUR @ 130%           55.4
```

Figure 9.10 Standard calculation sheet.

PART NO. A-111-222 R.L. QQ OPER NO. 000012

OPER. DSCRP. DRILL HOLES IN PLATE

PART NAME FLAT PLATE	COST CENTER 0000123
ISSUE NO. 1	MACH. CENTER 001111
DATE 1-29-79	ASSET NO. 000135
APPLICATOR WMY	CREW/MACH. 1/1
PLANT 02	

STANDARD CALCULATION

TYPE OF WORK	ELEMENTAL TIME	PERCENT ALLOWANCE	ALLOWANCE TIME	STANDARD TIME
EXTERNAL MANUAL	990.	15+4	188.	1178.
INTERNAL	(220.)			
PROCESS TIME	806.	30+15	363.	1169.
STANDARD(TMU/CYCLE)	1796.		551.	2347.
PIECES PER CYCLE	1			
STANDARD (HOURS PER PIECE)				0.023
(PIECES PER HOUR @ 130%)				55.4

Figure 9.11 Time calculation sheet.

STEP		FREQ	INTERNAL TMU	EXTERNAL TMU	LOC #
1		1.00		500.	134
2		1.00		50.	44
3	MACHINE OPERATION	1.00		806.	
4	INTERNAL TO STEP 3	1.00	220.		46
5		1.00		440.	136
MANUAL TIME(TMU)			220.	990.	
ACTUAL PROCESS TIME(TMU)			0.	806.	

Storage of Standards: Time Standards Database Program

Just as a suboperation is filed in a database by many categories, so is a completed time standard. The final time standard, along with the method instruction sheet and the standard calculation sheet, is filed in the standards database using the categories that appear in its heading (see Fig. 9.10) or any other categories so desired (see Fig. 9.12). Therefore, the standard can be retrieved by any one or any combination of the following:

Part number
Operation number
Operation description
Part name
Cost center
Machine center
Component classification number
Plant
Applicator
Date
Any other specific categories desired

The filing and retrieval of final time standards occurs in exactly the same manner as it does for suboperation data. That is, a search can be conducted by any one

Figure 9.12 MOST computer systems filing categories for time standards.

- PRODUCT / SUBASSEMBLY / PART NUMBER
- PRODUCT / SUBASSEMBLY / PART NAME
- COMPONENT CLASSIFICATION NUMBER

- PLANT NUMBER
- DEPARTMENT NUMBER
- COST CENTER NUMBER
- WORK CENTER NUMBER

- BILL OF MATERIAL NUMBER
- ROUTE SHEET NUMBER
- OPERATION NUMBER

- OPERATION NAME

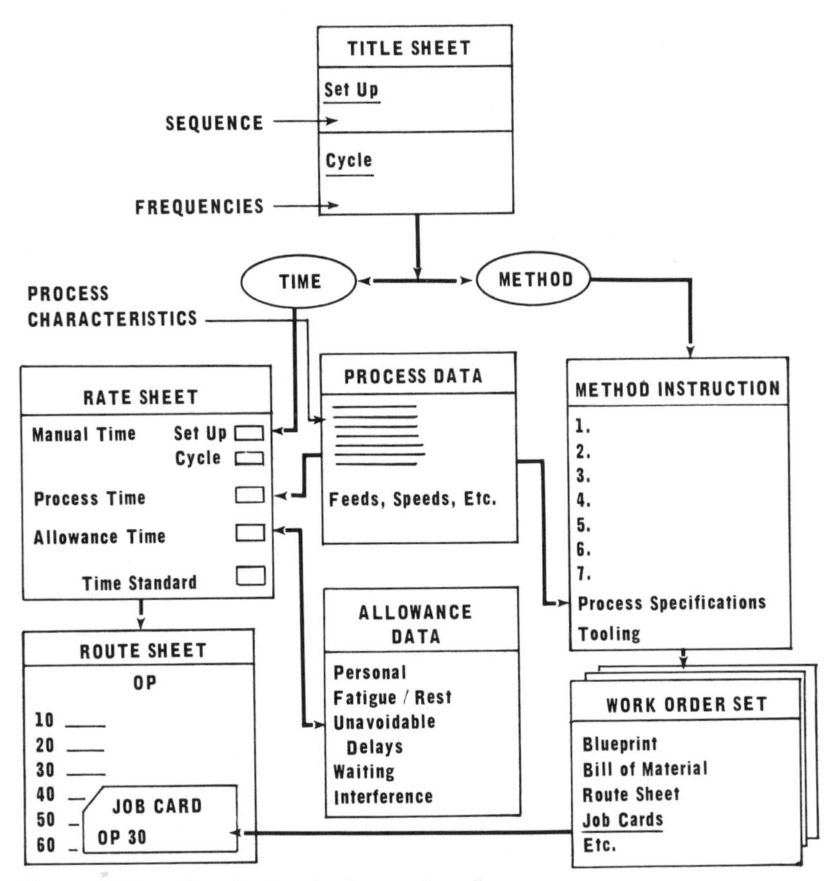

Figure 9.13 Standards calculation data flow.

or any combination of categories. When a time standard is filed, links to the title sheet and its components are established. If the standard is a new issue, comparisons can be made with previous issues, and decision rules for changes in incentive payments can be established according to a company's contract. Additionally, all operations under one number may be grouped on a route sheet with all information being fed to other production systems. Fig. 9.13 illustrates the complete data flow in the standards setting process.

Mass Update of Standards

As seen from the description of MOST Computer Systems, the database programs provide a complete linkage between workplaces, suboperation data, Title Sheets, and final time standards. These vital links provide the basis for an automatic update of time standards based upon changes in any of the basic data elements: workplace, method, or Title Sheet.

Too often in the industrial engineering application of standards, updating poses a problem because of difficulties in finding all standards affected by a change. Even if they can be found, usually a massive clerical task accompanies all changes. Because of these difficulties, "small" changes in the workplace or method go unrecorded. The cumulative effect of this procedure, however, leads to inaccurate time standards, sometimes resulting in a badly deteriorated incentive plan or a product costing completely off target.

The Mass Update program solves these problems. The "where used" option allows the user to query the database for all occurrences of standards dependent upon a basic data element which should be changed. This location feature results in a listing of all standards that would be affected by the change. The user then has an opportunity to edit the list for removal of standards that should not be changed.

The computer can then assume the clerical function of actually updating all the standards automatically, based on the changes in basic data supplied by the appropriate personnel. Since mass changes to the active standards will occur, this has to be a privileged feature, only available to the specified individual(s).

Probably the most exciting is the simulation of possible changes in methods or layouts as a response to the question "what if?" This simulation feature will open new doors for the industrial engineer in striving for improvement.

Once a substitution is made or suggested for a specified Title Sheet, the "where used" list is obtained, and the Mass Update command can then be issued to change the standards. Depending upon contract provisions, appropriate decision rules can be built into the change procedure. For example, if the standards should change only when a difference of plus or minus 5% occurs, this rule can be applied.

In essence, the Mass Update function is a valuable addition to the editing features in MOST Computer Systems. If a change affects only one or two standards, the editing feature should be used. But, when several standards are affected by proposed or mandated changes, the automatic facility of Mass Update is a necessary feature for keeping time standards current and accurate. The resources required for standards maintenance can be reduced by as much as 80 to 90% compared to a manual system.

Supplementary Programs

The Machining Data Program

The primary purpose of the speeds and feeds program is to calculate process times for machine operations such as drilling, milling, turning, gear cutting, and grinding.

The program determines the ideal speeds and feeds by using values recommended in the *Machining Data Handbook* (Metcut Research Associates, Inc., 1974). The user also has the choice of providing other source data, if desired.

In selecting speeds and feeds, consideration is given to material used, tool, machine specifications, dimensions of raw and finished workpiece, and so on.

Since the ideal speeds and feeds may not be available due to machine limitations, the program allows the user to select alternative speeds and feeds. This can be done automatically because the users' individual machine specifications have been included in the data file.

Outputs of the program are the speeds in rpm, feeds, recommended tool grade for the material being machined, the process time and allowances inherent in the maching operation such as lead-on and overrun, and power utilization.

For turning and milling operations, the program also calculates the number of cuts. All speeds and feeds are based on a tool life of 1 hour. Figure 9.14 is an example of input for a drilling operation. Figure 9.15 shows the resulting output.

Machining Data Output Interpretation For mild carbon steel 1020 material with a BHN 125,175 running on a ratio multiplier drill RI4125 (horsepower of 10), five ¾-in. diameter, 1.5-in. deep holes were drilled. The output, shown in Figure 9.15, gives a recommended surface speed of 90 FPM and a feed of 0.012 IPR. After this is compared to what is available on the machine, the program selected a rpm of 495 (both the setting and actual are the same in this case) and a feed rate of 5.9079 IPM. The total time for the drilling and rapid travel are 806 TMU. The frequency for tool change is 1/236.

Figure 9.14 Machining program data input.

```
        MACHINE SPEEDS AND FEEDS PROGRAM


PART NO ?A-111-222

OPERATION NUMBER ?000012

STARTING FROM A SAVED FILE <Y,N>N

MATERIAL CODE ? 1020

  HARDNESS = 125,175
MACHINE NUMBER ?RI4125

TYPE OF DRILLING<H=HELP> ? D

DIAMETER ? 3/4

DEPTH OF HOLE ? 1.5

BLIND =B,THROUGH=T ? T

NUMBER OF HOLES ?5

EFFICIENCY OF MOTOR ? 1.
```

```
**************************************************************************
     RECOMMENDED SPEEDS AND FEEDS FOR DRILLING

     SPEED=  90 FPM
     RPM=  458
     FEED= 0.0120 IPR
     TOOL GRADE-M10,M7,M1
     MACHINE TIME=                    454
     COOLING SYSTEM AVAILABLE

     DATE = 29-JAN-79
    PART NUMBER=A-111-222        OPERATION NO=000012
    MACHINE NO.=RI4125          MATERIAL CODE=1020
                                  SET       ACTUAL                    TC.
 STEP  TITLE            HOLES    RPM        RPM  FEED RATE   TMU
  1D   0.7500TD  1.5000     5    495        495  5.9079      592      236
                                      RAPID TRAVEL TIME =    214
                                            TOTAL TIME =     806
```

Figure 9.15 Machining program output.

The Welding Program

The welding program calculates the process time (arc time) for arc-welding operations using either rod or wire electrodes. The calculations are based on filed geometric joint descriptions, electrode characteristics and methods (electrodes, amperage, wire speed, etc.). The output is an arc time and summary report. Figure 9.16 is a sample arc time report for the fillet joint shown in Figure 9.17. The joint and method identifications are given by the user, along with the weld length. These three items, plus the number of passes (from the method description), are shown in the arc time report heading line. The electrode identification, voltage, amperage for rod electrode, and wire speed for wire electrode are shown (from the method description), along with the calculated travel speed, frequency of electrode change (EL. CH.), and arc time. The travel speed is the rate at which the rod or wire should move along the weld; the electrode change frequency is applied as a frequency to the MOST calculation for the manual operation performed when changing such an electrode.

The Line-Balancing Program

The line-balancing program is designed to solve the problem of assigning work elements to assembly line stations in order to minimize the balance delay.

The program can assign elements to a station either for a desired cycle time or for a designated number of stations. The program seeks to assign the elements to each work station in a manner that minimizes the number of stations required (if cycle time is used) or the cycle time (if a number of stations are assigned). This is accomplished quickly and without violating the technological constraints governing the order in which these elements can be performed. Output is in the form of the three best solutions in terms of minimizing balance delay. The user then has the option of moving elements from one station to the other if desired. The program will then recalculate the balance delay. Figure 9.18

JOINT ID : FILLET-1/4"

METHOD ID : 1-PASS-9018-3/16

LENGTH (IN) : 10

==

STEP 1: FILLET-1/4", 10.0 IN LONG, IN 1 PASSES (1-PASS-9018-3/16)

ELECTRODE	VOLT	AMPS	WIRE SP.	TRAVEL SP.	EL.CH.	ARC TIME (TMU)
9018-3/16	24	250		6.5	0.9138	2560.

==

Figure 9.16 Welding arc time report.

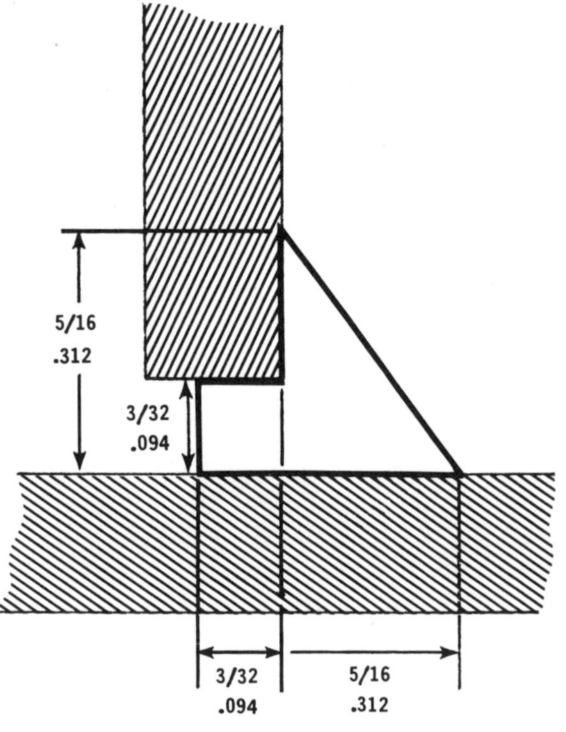

Figure 9.17 Fillet joint.

```
ENTER NO. OF STATIONS YOU WISH TO DESIGNATE ? 4

 NO. OF STATIONS DESIGNATED IS      4


LINE BALANCING RUN NO.     1

CYCLE TIME =                        3940.
NUMBER OF STATIONS =                  4
NUMBER OF ITERATIONS =               2
BALANCE DELAY =                     3.9 PERCENT

    WANT DETAIL ANALYSIS ? Y

STA    EL     LOC      TMU NOEL WKTM IDTM
  1
         1     0      200. ASSEMBLE SHEET METAL & HARDWOOD STOCK
        11     0      450. ROUGH SAW HANDLES AND TRAY BASES
        12     0      930. SQUARE AND TAPER HANDLES
        14     0      640. FORM TRAY BASES
        13     0      700. FORM HAND GRIPS
        15     0      320. DRILL HANDLES AND TRAY BASES
         2     0      700. TRIM SHEET METAL FOR TRAY,LEGS,CENTERBRACE,NOSE GUARD
                        7   3940.0      0.0


  2
        16     0     1450. SAND HANDLES AND TRAY BASES
        18     0      680. FINISH TRAY BASES
        17     0     1120. VARNISH HANDLES
        20     0      480. BOLT WHEEL BEARINGS ON HANDLES
                        4   3730.0    210.0


  3
         5     0      150. STAMP NOSE GUARD
         4     0      250. STAMP LEGS & CENTERBRACE
         7     0      850. DRILL LEGS,CENTERBRACE,NOSE GUARD
         9     0      120. CRIMP SIDES OF NOSE GUARD
         3     0      550. FORM TRAY
         6     0     1400. BEAD TRAY EDGES
         8     0      240. DRILL TRAY
                        7   3560.0    380.0


  4
        10     0     1250. PAINT TRAY,LEGS,WHEEL,CENTERBRACE,NOSE GUARD
        21     0      720. MOUNT TIRE ON WHEEL
        19     0      520. ASSEMBLE LEGS AND CENTERBRACE
        22     0      600. ASSEMBLE WHEEL,LEGS,HANDLES & NOSE GUARD
        23     0      830. BOLT TRAY TO BASES AND HANDLES
                        5   3920.0     20.0

LINE BALANCING RUN NO.     2

CYCLE TIME =                        4200.
NUMBER OF STATIONS =                  4
NUMBER OF ITERATIONS =               2
BALANCE DELAY =                     9.8 PERCENT

    WANT DETAIL ANALYSIS ? N

LINE BALANCING RUN NO.     3

CYCLE TIME =                        4410.
NUMBER OF STATIONS =                  4
NUMBER OF ITERATIONS =               2
BALANCE DELAY =                    14.1 PERCENT

    WANT DETAIL ANALYSIS ? N

LINE BALANCING RUN NO.     4

CYCLE TIME =                        4030.
NUMBER OF STATIONS =                  4
NUMBER OF ITERATIONS =               2
BALANCE DELAY =                     6.0 PERCENT

    WANT DETAIL ANALYSIS ? N
```

Figure 9.18 Line balancing solutions.

provides an example of a line-balancing solution for a wheelbarrow assembly line.

The System

The MOST Computer Systems is written for a Digital Equipment Corporation PDP-11 minicomputer to be located for direct on-line access in the industrial engineering department. The programs, which utilize random access techniques, run under a minimum 32,000 words of memory, with additional requirements determined by the number of simultaneous users. Disk storage capacity depends totally upon the amount of data to be stored (total number of time standards). MOST is also available for IBM main frames having TSO.

The programs run completely "on-line" in a time-shared environment. CRTs are hardwired directly to the computer or can be connected by dedicated telephone lines. Printers are used to obtain hard copies of the data displayed on the CRT. Finally, a word-processing system is available. Its function is to relieve the engineer of the task of developing, from scratch, work management manuals (Chap. 8). The word processing system can store a database of work management manuals that can be quickly retrieved and edited, eliminating the time-consuming and costly function of rewriting and retyping manuals. The result is complete standards development and documentation under the guidance and control of the computer. This produces accurate and consistently applied time standards.

Summary

As MOST Computer Systems is based on the MOST Work Measurement System and the MOST Applications System, it shares their features, such as, consistency of application and documentation, a high degree of accuracy, uniformity among applicators, complete traceability of all backup data, and ease in learning and application. In addition to these features, MOST Computer Systems has some advantages of its own.

- It eliminates routine work, nearly all paperwork, and the use of a stopwatch.
- It is 2 to 5 times faster than the Manual MOST System.
- Work area layouts are an integral part of the system and remain linked to the analysis throughout.
- Supplementary programs provide accurate process times tailored to company machines and welding processes.
- Through the editing process, changes in workshop conditions are easily implemented and documented and the standards automatically adjusted and updated.

- The editors provide an opportunity for detailed methods analysis through simulation.
- The filing and retrieval system opens a host of possibilities for data organization, sharing data among plants or areas of a single plant, mass updating, and formulating of prototype work areas, suboperation data, and final time standards.
- It provides operator method instruction and route sheets as byproducts of the time standard calculation process.
- The time standards can be easily linked into larger host computers for use by payroll, production control, forecasting, scheduling, costing, and other programs.

In essence, MOST Computer Systems is designed to assist the industrial engineer to become more productive on the job, with more time for concentration on new methods to increase manufacturing productivity. Because it is simple to learn and apply, and can be easily related to the manual process, MOST Computer Systems can readily gain acceptance by union officials and workers, as well as management.

Simplicity, accessibility, consistency, and speed are the major features that make MOST Computer Systems an excellent tool for the industrial engineer, to increase self-productivity and make an important contribution to both departmental and company profitability.

10 In Summary

In summary, the MOST Systems family (Fig. 10.1), as presented in this text, provides the industrial engineer the tools with which to measure, document, and control manufacturing methods and costs. This family of techniques and systems is based on the strong support of its oldest member, the MOST Work Measurement Systems. Realizing that a higher level of work measurement technique requires a higher level of application procedure, the MOST Application Systems were developed with specific guidelines for the major industries of general manufacturing, textiles, shipbuilding, and for clerical operations, etc. The MOST Computer Systems contains both the above-named family members, and, as a result, it documents and guides the work measurement process from the establishment of the workplace through the development of the final time standard.

The MOST Systems family has grown significantly into a comprehensive system since its introduction a few short years ago. It can be applied manually or by the assistance of modern computer technology. The future will bring further sophistication, refinement, and development.

Significant Concepts

In review, it might be worthwhile to restate a few of the significant concepts upon which the MOST Systems are built.

The Sequence Model

With the sequence model rests the fundamental concept on which MOST originally was built. Because of the development of the sequence model, the

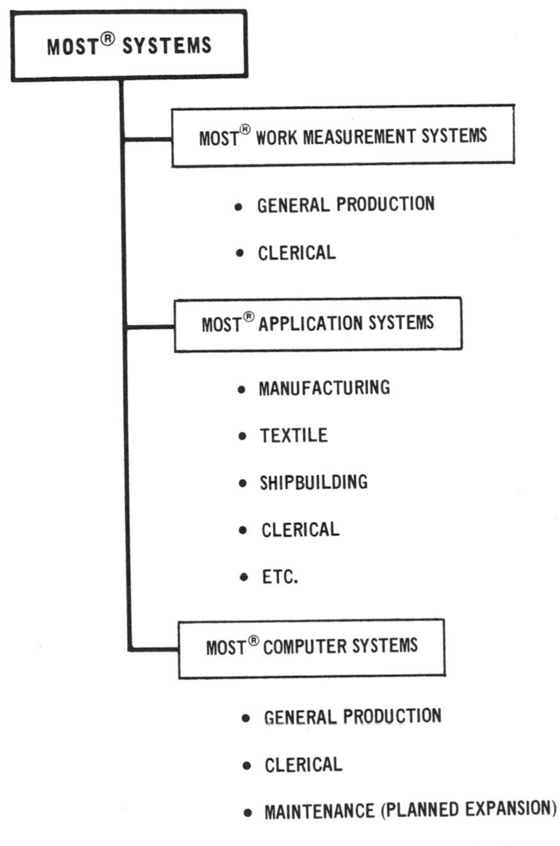

Figure 10.1 MOST Systems family.

analyst's focus is shifted from the operator's body movements to the movement of objects. This provides a larger unit with which to work and a resulting clearer, more understandable method description. Because of the sequence model, the analyst is forced to consider all the subactivities possible to move an object. As a result, more analyst consistency and less application error should be seen.

Because of the sequence model, the lowest level of meaningful data should now appear as the complete movement of an object, not, for example, only the movement of the hand to reach for an object. It is from this higher basic level that work is analyzed and "suboperation data units" result.

Suboperation Data

With MOST, it is both impractical and highly inefficient to break work (operations) down into data units any smaller than a complete and logical segment of work, an "activity." In other words, there is no need to develop data units smaller than a logical sequence of method steps as analyzed on a MOST calculation form. Data units, the size of a completed MOST calculation, can be more easily understood and more efficiently filed and retrieved than smaller data elements. And, more importantly, they can be revised and/or updated on one high level (the MOST calculation form), not through a series of standard data building block levels. Using MOST, the analyst should look at the entire operation (start at the top) and if necessary break that operation *down* into logical activities or suboperations. This "top-down" approach to data development is completely counter to previous standard data building block approaches, where standard data units are *built up* from basic elements through a series of levels. The high-level suboperation data unit is the level at which analyses are edited and/or updated, producing times specific to particular operations with little additional analyst time or effort required.

The Statistical Foundation

The MOST Work Measurement System is based on the fundamental statistical standard deviation concept. It is through the application of "engineered deviations" that the system gains not only its speed, easy application, and accuracy, but also an outstanding consistency. The MOST Work Measurement System as described in this book was *designed* to produce a predetermined level of accuracy. Other systems were created and then their accuracy was determined. The statistically based deviations on which MOST is built provide accurate and consistent results throughout its application. With MOST, the deviations—the index ranges the index numbers represent—are "engineered inaccuracies," they do not occur haphazardly across the work measurement spectrum. Therefore, industrial engineers always know the system accuracy and the confidence level with which they are working.

Appendix A provides detailed information regarding the statistical foundation of MOST and the development of the index numbers. Reading it will give you an appreciation of the system design, an understanding of the support that underlies the MOST Systems, and confidence in MOST analyses.

MOST Is a Method-Based System

As with all work measurement techniques, the time values that result from performing a measurement should always be based on a specific and well-engineered method. The prime job of the analyst, then, is to properly determine the representative method of an operation or suboperation. The second step, the analysis, is then easy to do when using the MOST Systems. From an engineering point of view, any deviation between the calculated time and the "actual time" lies between the "engineered method" and the method actually performed by the operator. The problem then is not in engineering but in education. If the operator is not following the engineered method, the question must be asked, "Was the operator properly trained?" If not, why even bother to "engineer" the job? Therefore, utilize the method description section [Chap. 7, Fig. 7.1, section (4)] portion of the MOST calculation form to train the operators to use the engineered method.

Communication—"Closing the Gap"

How do you close the gap between the industrial engineer and the supervisor, foreperson and worker in the shop? Use the method sensitivity of MOST to do away with nonproductive efforts. Use the method description section of the MOST calculation form to train the operators. Use the communicative powers of this "easy to learn and understand" system to inform workers and their representatives of the method and actual calculation if required. MOST now provides the communication link (a common and easily understood language) that heretofore has been lacking in work measurement systems, through which the worker and the engineer can speak of and understand the work both are doing. The result will be the early alleviation of problems or potential problems at the analysis and installation level where they actually occur. Having achieved a desired level of understanding between those directly involved, the organization can function smoothly, allowing common productivity and profit goals to be realized.

appendix **A**

Theory

With existing predetermined motion time systems (PMTS), such as MTM, an "exact" method description must be recorded, i.e., basic motions must be expressed in terms of distinct types, distances, weights, and other such variables. The result is a very detailed description of a method, but probably not the exact method actually performed in the majority of cases. Variations are inherent in any operation. The operator may follow a different motion pattern due to a lack of instructions or because of the variable nature of the operation, e.g., reach and move distances may not be the same throughout an operation as the detailed description implies. The analyst must predict what methods will be used or describe what is perceived to be an average or representative method for the operation. When distances are averaged, for example, as in reaching into a bin with parts at different distances from the operator, the expected accuracy of the detailed system is not achieved. Why describe a detailed method from a table of "exact" values, when, in actuality, the operator will follow a method that varies from one occurrence to the next? There are, of course, situations when this detailed approach is indeed appropriate, such as for highly repetitive, short-cycled operations performed at workplaces designed to minimize any such variations. But, in terms of cost, this exacting method analysis required by existing PMTS work measurement techniques often seems unnecessary, if not impractical. In fact, there is even some question as to the exactness of these systems given the presence of this inherent variability in work methods.

In the design of MOST, it was recognized that these variations or deviations could be easily compensated for by using basic statistical principles. But, most importantly, by using these same procedures, it was also found that it was possible to greatly simplify the work measurement itself while retaining a high level of accuracy. In other words, this inherent variation in work methods actually can be used to an advantage in developing a more simplified work measurement technique, with resulting accuracy surprisingly close to systems such as basic MTM.

Accuracy of a Predetermined Motion Time System

Can a work measurement technique produce a time that is exact? The answer to this question is *no*, because:

- Work is performed by humans and is, therefore, *variable* from one human to another—no two individuals will accomplish "exactly" the same amount of work over a specified amount of time.
- Time standards (the results of work measurement) represent average times. They reflect the time of an *average* worker, of *average* skill, working at an *average* or normal pace, under *average* conditions.

Therefore, since no work measurement technique is exact, all have as either part of their original system design or as a result of their original design, a *balancing time.* By definition, balancing time is the time needed for the system's desired level of accuracy to be attained. In other words, a certain amount of time, generated by completing analyzed work, must be accumulated when using a system, before the accuracy of that system can be guaranteed to a specific level of confidence.

The statistical phenomenon that occurs *during* that balancing time is called the *balancing effect.* The balancing effect is what causes the desired level of accuracy of a system to be attained. Therefore, the balancing time is *when* the system's accuracy is attained and the balancing effect is *how* it is attained. By definition, balancing effect is the combination of individual deviations for a smaller total deviation. Deviation can be defined as the difference between the "true" time it takes for a task and the time the work measurement technique "allows."

Balancing effect is based on the statistical principle that the variance of a sum of independent variates equals the sum of the individual variances; that is, as individual elements are added, the total percentage deviation will become less than the individual percentage deviations. This is because these deviations balance each other due to some being above the true time and others below. In the final result, the total relative accuracy is better than the accuracy for the individual measurements.

Admittedly, the true time to perform a job is unknown, therefore, in designing the MOST Systems, and when speaking of its accuracy, MOST is compared to a "true time value" as determined by the proper application of MTM-1.

MOST System Design

Unlike MTM, whose balancing time is a result of its system design, the balancing time of MOST was a calculated part of its system design. To better understand how MOST produces accurate results, a look at the system design is necessary.

When MOST was originally conceived, the decision was made that a balancing time of approximately 2 minutes would be desirable. It was reasoned that substantial simplicity in system design and application could be achieved with only a moderate reduction in accuracy. The system, therefore, was constructed to have a consistent balancing time of approximately 2 minutes (3000 TMU was the target value for the original calculation.)

The actual balancing time for MOST was determined by Dr. William D. Brinckloe of the University of Pittsburgh* and was found to be 3235 TMU. In the next section of this chapter, which covers the actual system construction, all calculations and index ranges are based on the balancing time of 3235 TMU. Therefore, the accuracy of MOST is based on a balancing time of 3235 TMU, or approximately 2 minutes. That is, measurements totaling 3235 TMU or more will be accurate to within ±5% of a true time value with a confidence level of 95%. This does not mean, however, that MOST can never be used to measure shorter activities. It should again be pointed out that accuracy in the *final* result (the standards level) is the deciding factor. In other words, the minimum condition of 3235 TMU applies only to the final result (the standard) and not to the individual measurement. Also, the practical accuracy, that is, the results of application, of MOST and of MTM are extremely close.

MOST Interval Groupings

Predetermined motion time systems, such as MTM, are constructed by determining the time duration of conveniently selected basic motions. In contrast, MOST starts with the construction of time intervals *based on a stated balancing time* (3235 TMU) and thereafter determines what motion patterns fall within each time interval. The MOST Systems, therefore, provides a consistent balancing time throughout.

Influential in the construction of the MOST Systems time intervals was the establishment of the following objectives:

*The theoretical system accuracy of MTM-1, MTM-2, and MOST is discussed in *Comparative Precision of MTM-1, MTM-2 and MOST,* University Research Institute, June 1975.

1. The mean value for each time interval will be a whole number and also a multiple of 10.
2. The time intervals will cover a continuous time scale with neither gaps nor excessive overlap.

The MOST time intervals were then calculated from the statistical formula for allowed deviation:

$$a = \pm r_{T_B} \sqrt{T_B \times t}$$

where a = allowed deviation from interval mean in ± TMU
$\pm r_{T_B}$ = accuracy of ±5% for balancing time
T_B = established balancing time of 3235 TMU
t = interval mean in TMU (a whole number and also a multiple of 10)

Table A.1 is the result of using this formula with appropriate values.

The formula stated above assumes a conventional normal distribution. If a uniform distribution is assumed, the formula will then become:

$$a = \pm r_{T_B} \sqrt{T_B \times t \times 0.878}$$

The differences in Table A.1 from using this second formula are quite minor and do not affect the construction of the MOST index number table (Table A.2) or the location of the boundaries between index numbers.

By placing the values from Table A.1 on a linear scale and drawing half-circles, which represent the calculated allowed deviation range of each time interval, the graphical representation shown in Figure A.1, is produced for the first five intervals or index groups. Note that since the interval means were adjusted to be divisible by 10, the application of MOST is simplified by eliminating zeros, thus creating a series of index numbers (circled) statistically representing each time interval.

Table A.1

INTERVAL MEAN TMU	DEVIATION TMU	INTERVAL LOWER LIMIT TMU	INTERVAL UPPER LIMIT TMU
10	9	1	19
30	16	14	46
60	22	38	82
100	28	72	128
160	36	124	196

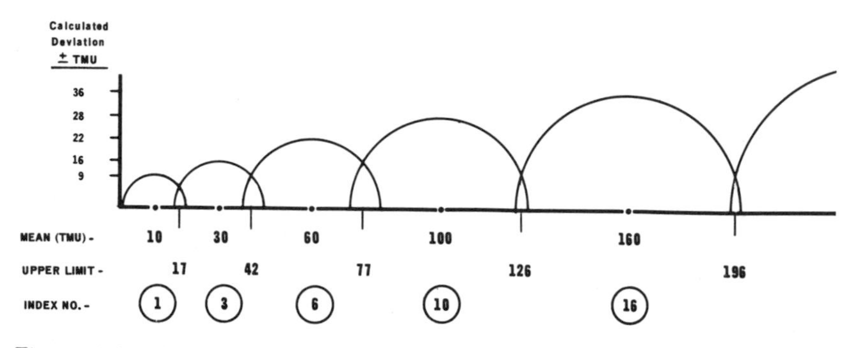

Figure A.1 Construction of MOST interval groupings.

Table A.2 MOST Index Series

INDEX NUMBER	INTERVAL MEAN TMU	MOST INTERVAL LIMITS TMU
0	0	0
1	10	1 - 17
3	30	18 - 42
6	60	43 - 77
10	100	78 - 126
16	160	127 - 196
24	240	197 - 277
32	320	278 - 366
42	420	367 - 476
54	540	477 - 601
67	670	602 - 736
81	810	737 - 881
96	960	882 - 1041
113	1130	1042 - 1216
131	1310	1217 - 1411
152	1520	1412 - 1621
173	1730	1622 - 1841
196	1960	1842 - 2076
220	2200	2077 - 2321
245	2450	2322 - 2571
270	2700	2572 - 2846
300	3000	2847 - 3146
330	3300	3147 - 3446

The diagram shows that adjacent half-circles overlap slightly. The overlaps have median values of 17, 42, 77, 126, and 196 TMU. These are the upper limits of the first five MOST index ranges. The calculation procedure was continued to determine the values shown in Table A.2.

It is within these statistically calculated index ranges that the variation inherent in the majority of working situations will be absorbed. The system balancing time and the balancing effect will produce accurate results.

Backup Data

The MOST time intervals described in the previous section serve as the basis for all parameter index values. Motion patterns are analyzed with MTM-1 or MTM-2 and assigned index numbers corresponding to the time interval into which the detailed analysis falls. The most frequently occurring of these motion patterns are listed on data cards under appropriate sequence model parameters and comprise the variants for the various sequence model subactivities defined in earlier chapters.

Each of these motion patterns (variants), with its corresponding index number, is referred to as a "parameter index value." For example, MTM analyses for "gaining control of an object requiring disengage" fall within the time interval 18 to 42 TMU. From Table A.2, this translates to index number 3. Therefore, Gain Control with Disengage is represented in the sequence model by the parameter index value G_3. For every value on the MOST data cards, there are corresponding MTM-1 or MTM-2 analyses cataloged in a backup data manual.

Applicator Deviations

The total accuracy of any work measurement technique is dependent on both the system deviation and the applicator deviation. While system deviation can be determined statistically, the deviations present, because of applicator error, must be determined empirically. Applicator deviations will vary with individuals, depending largely on the amount of training and experience possessed by each analyst.

One of the basic assumptions concerning the accuracy of a work measurement system is that a fine breakdown of motion variables contributes to the reduction of system deviation. While this is no doubt true, there is at the same time a greater tendency for the applicator, either through inexperience, misjudgment, or carelessness, to make errors in the selection of the correct time value. As a result, the systems taken to be the most precise (i.e., those with the finest subdivisions of motions) are the most susceptible to applicator errors.

Applicator deviations can be influenced by the way in which motions are classified within the system. With existing PMTS systems, as many as four variables must be considered when selecting time values. For example, MTM-1

Reach time values are classified by distance, case, and type and Move time values by distance, case, type, and weight. Obviously, the more variables that must be considered by the analyst, the more likely the possibility for applicator error. Index values in MOST were designed to contain but one variable. To select the proper index value for Action Distance, only the distance is considered.

Perhaps the most frequent type of applicator error is that of carelessly omitting a motion from a motion pattern or erroneously including a motion that does not occur. This problem is virtually eliminated in MOST with the aid of the preprinted sequence models. During the analysis procedure, the applicator's attention is focused on each sequence model parameter as the calculation sheet is filled out.

Surprisingly, very little research has been done on evaluating applicator deviations in the various work measurement systems available today. This is unfortunate, since applicator error probably influences the total accuracy as much as or more than the system error. However, there is analytical evidence to indicate that the total accuracy of MOST is influenced to a lesser degree by applicator deviations than existing predetermined motion time systems. It is felt that the loss in system accuracy (the larger balancing time of MOST) could be compensated for by a reduction in applicator error, thus pulling the more detailed system and the more economical system together into an area of comparable total accuracy.

Accuracy of Time Standards

Having reviewed the theory and construction of the MOST Work Measurement technique, let us now look at this same theory extended to a higher level, that of analyzing operations or suboperations to establish a time standard.

Accuracy of Work Measurement

A condition is said to be "accurate" when it conforms exactly to an accepted standard; i.e., the condition falls within acceptable tolerance limits. Accuracy, then, is a relative concept, relative to an accepted standard. To a carpenter, accuracy is usually expressed in inches or eighths of an inch, but to the machinist may be expressed in thousandths of an inch. The physicist deals with even smaller tolerances. What about the work analyst? What is the "accepted standard" for work measurement accuracy. Should operation times be accurate within thousandths of a second or would plus or minus 1 day be acceptable?

Obviously, both these conditions are inappropriate. In one case, the measurement would be extremely difficult to obtain, and in the other, the results would probably be meaningless. The fact is that accuracy requirements have very little to do with work measurement. The main consideration is of an economic nature. If it costs thousands of dollars to develop time standards for an operation that seldom occurs, we will be satisfied with a rough estimate or even a guess. On the other hand, if substantial economic benefits can be realized from a detailed analysis providing more "exact" times, these studies may well be worth the cost. So, the question of work measurement accuracy must first of all be answered in terms of the cost involved to achieve a certain level of accuracy.

Theoretically, the accuracy of any work measurement system can be defined as the deviation or percent difference between the analyzed time and the true time. In terms of both the total standard and individual analysis deviation, the accuracy (in percent) of MOST can be defined as follows.

Unit Relative Deviation

$$r_t = \frac{\text{MOST} - \text{true}}{\text{true}} \times 100 = \frac{a}{t} \times 100$$

where a = the deviation from the actual unit time in TMU
t = the actual unit time in TMU

Total Relative Deviation

$$r_T = \frac{\text{MOST} - \text{true}}{\text{true}} \times 100 = \frac{A}{T} \times 100$$

where A = the deviation from the total true time in TMU
T = the total true time in TMU

As stated earlier, the true time, of course, is unknown. Therefore, we must use some standard for comparison. MOST will be compared to a "true time value" or the time which results from properly applying a recognized work measurement technique such as MTM-1.

The accuracy of a work measurement system is influenced by four factors, which, when assembled as a formula, explain the total relative deviation theory. These four factors are:

1. The *level of accuracy desired* in the final result. This will depend on the planned use of the time standard, such as incentive payment calculations (individual or group), machine loading, and product costing.
2. The *time period* over which the desired level of accuracy must be attained. Do we want these *time standards* to achieve the desired level of accuracy on a 1-day basis, or will accuracy based on the 40-hour week be sufficient? This period will be referred to as the calculation period, leveling period, or balancing time. *Note:* We are now discussing the balancing time for time standards calculation, *not* the balancing time of the work measurement technique used to determine the standard times.
3. The *degree of repetitiveness* of the suboperation being measured. That is, how many times does the suboperation occur during the calçulation period?
4. The *duration of the suboperation* being measured.

These four factors are mathematically represented by the following statistical formula used to calculate "allowed deviation." The formula is a direct derivation of the well-known expression for the standard deviations:

$r^2 t = \text{constant}$

Each formula-variable definition is followed by a number referencing it to one of the four factors mentioned above.

$$r_t = \pm r_T \sqrt{\frac{T}{nt}}$$

where r_t = the measured suboperation's allowed deviation in percent
r_T = the total allowed deviation in percent (1)
T = the total time, i.e., the calculation period or balancing time (2)
n = the suboperation's occurrence frequency over the calculation period (3)
t = the suboperation's measured time (4)

Using the formula presented above, the allowed deviation of a suboperation is calculated under two different conditions in the following example.

Example: A typed report required 0.25 hours to perform according to a work measurement analysis. How accurate must this analysis be (i.e., what is the allowed deviation) if the time will be used for setting incentive rates where standards are expected to be within ±5% for a 40-hour pay period?

Case 1 The report is typed by a receptionist only twice a day.

$r_T = \pm 5\%$ $\qquad r_t = \pm r_T \sqrt{\frac{T}{nt}}$

T = 40 hours

n = 2 / day × 5 days = 10 $\qquad r_t = \pm 5 \sqrt{\frac{40}{10 \times 0.25}}$

t = 0.25 hours $\qquad r_t = \pm 20\%$

Case 2 The same report is typed continuously during the 40-hour calculation period by a word processor.

$r_T = \pm 5\%$

T = 40 hours $\qquad r_t = \pm r_T \sqrt{\frac{T}{nt}}$

$n = \frac{40 \text{ hours}}{0.25 \text{ hours}} = 160$ $\qquad r_t = \pm 5 \sqrt{\frac{40}{160 \times 0.25}}$

t = 0.25 hours $\qquad r_t = \pm 5\%$

In Case 1, a deviation of ± 20% (0.25 ± .05 hours, or between 12 and 18 minutes) can be allowed for the standard time established since the job occurs infrequently and makes up only a small part of the receptionist's total productive time (Fig. A.2). In Case 2 the word processor types reports during the entire week,

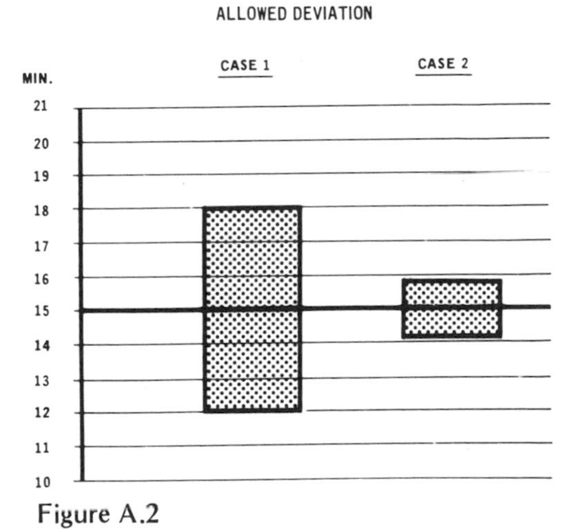

Figure A.2

causing the relative deviation to remain constant. The allowed deviation in this case is ±5%, the same as the requirement for the 40-hour calculation period. This relates to an allowed time of 0.25 ± 0.0125 hours or a "standard time" for the task anywhere from 14.25 to 15.75 minutes (Fig. A.2). These statistically calculated allowed deviations guarantee with 95% confidence that a calculated time which falls within the allowed deviation range will, over the calculation period, produce results within ±5% of a true time value.

Therefore, frequently occurring suboperations, or those under restraints of a shorter leveling or calculation period, have a very limited range of allowed deviation associated with each and as a result can absorb very little in the way of method variations in the operation. For example, the one analysis for Case 2 will statistically represent the typing of other reports, or variations of the one analyzed, which fall in a range of from 14.25 to 15.75 minutes. On the other hand, the analysis for Case 1, with its lower frequency of occurrence, will statistically represent a wider range of variation, for example, the typing of any report or letter taking from 12 to 18 minutes to prepare. The result of analysis at this higher level is statistically determined and leads to a reduced number of time standards. The calculation of fewer standards will obviously save time and effort for the industrial engineer or standards applicator. Chapter 8 addresses this subject.

Accuracy Test

The previous accuracy calculations were also based on the balancing time and balancing effect theories presented earlier. Balancing time is graphically illustrated in Figure A.3.

As the diagram in Figure A.3 shows, the desired level of accuracy of ±5% (r_T) is required to be reached as the sum of the individual measurements (t) approaches a certain point. The total time at this point is referred to as the balancing time (T), which in this case is 40 hours. A balancing time of 40 hours, as in this example, allows a wide margin for variation while setting the time of individual activities in most typical cases. Most of the predetermined motion time systems, including MOST, are capable of far more accuracy than this as a general rule.

As for the balancing effect, it can be tested by evaluating the deviation between the true times for different suboperations and the allowed times for these same suboperations. According to theory, the desired level of accuracy should be achieved at the calculated balancing time.

Table A.3 lists 10 time ranges covering 0.0 to 11.0 hours and the allowed time representing each of these ranges. In the middle columns of the table the maximum allowed deviation for each time is shown. All table values were determined from the allowed deviation formula for a 40-hour balancing period with a 95% confidence level.

To test the accuracy of this system, a series of random numbers should be generated to represent the "true times" to perform certain suboperations. How-

Figure A.3 Illustration of balancing time.

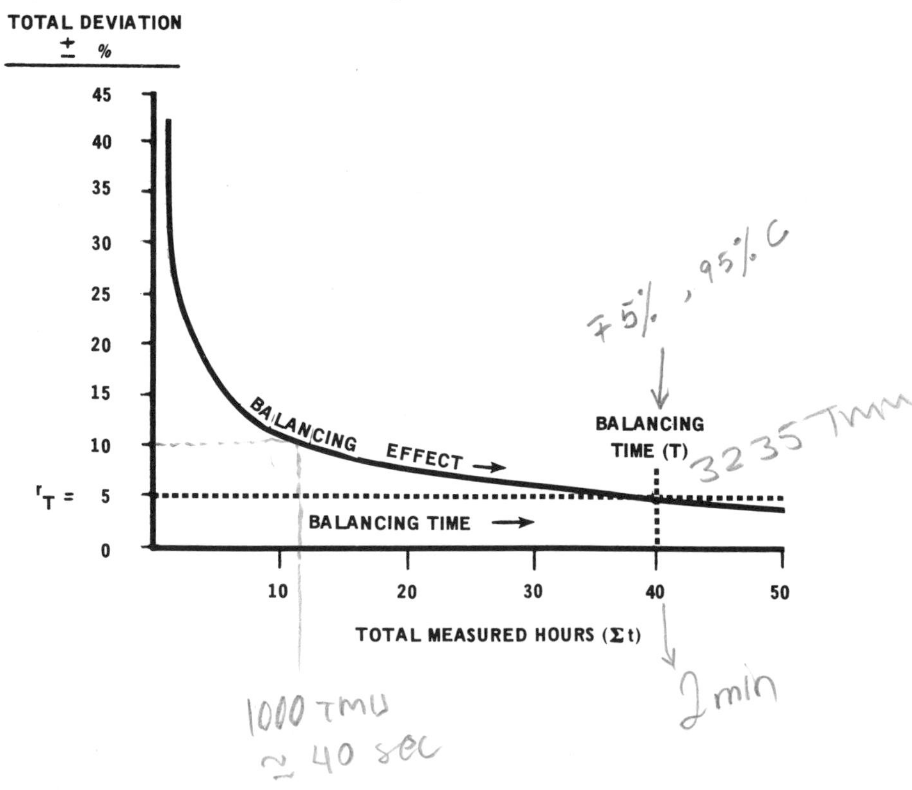

ever, for a simple demonstration, random numbers can be generated, for example from a telephone book. *Note:* It is recognized that a series of numbers generated from a telephone book may not necessarily be random.

The last two figures of the telephone number represent the "true" time for the suboperation in hours, with one decimal place. For example, the number 412-2375 would generate the true time of 7.5 hours.

After the true time has been established from the random number table or a telephone book, the next step is to select the appropriate allowed time based on the range into which the actual time falls. Using Table A.3, it can be seen that the time of 7.5 hours falls within the range 7.2 to 9.0 hours; therefore, a time of 8.1 hours would be allowed since it is the midpoint of the range. In order to evaluate the balancing effect theory, enough values must be chosen so that the total of the allowed times is at least 40 hours (the balancing time of Table A.3). This is necessary to ensure that the desired level of accuracy is achieved with a 95% confidence level. A complete test example is tabulated in Table A.4.

Notice that while individual deviations were as large as 20% in one instance, the total deviation was better than the ±5% specified. If we relate this example to a "real" work measurement situation, we can see that calculated deviations can be allowed in individual measurements without losing the level of accuracy desired in the final result for a calculation period (1 day, 1 week, etc.). This balancing principle plays an important role in the conceptual design of

Table A.3 Allowed Time Values for a 40-Hour Balancing Period

RANGE HOURS	MAXIMUM ALLOWED DEVIATION		ALLOWED TIME VALUES IN HOURS
	± HOURS	± %	
0.0 - 0.2	0.1	100	0.1
(0.2) - 0.6	0.2	50	0.4
(0.6) - 1.2	0.3	33	0.9
(1.2) - 2.0	0.4	25	1.6
(2.0) - 3.0	0.5	20	2.5
(3.0) - 4.2	0.6	17	3.6
(4.2) - 5.6	0.7	14	4.9
(5.6) - 7.2	0.8	13	6.4
(7.2) - 9.0	0.9	11	8.1
(9.0) - 11.0	1.0	10	10.0

MOST and the MOST Application System. The balancing effect is fairly well demonstrated in the above example. However, a single simulation is rather meaningless to prove the general significance of the balancing effect. Therefore, a computer program for a random simulation was written. Two runs of that program, each with a sample size of 100 simulations, showed that the average percentage error was well within ±5%—2.69% and 2.63%, respectively.

In each of the previous examples, the desired level of accuracy was specified to be ±5%, which is the generally accepted standard for industry. But what about the balancing time? The 40-hour balancing period may be sufficient for calculating incentive standards based on a 40-hour pay period, but hardly acceptable for a line-balancing calculation with cycle times in minutes. The use of the time standard is, therefore, a very important factor when considering the balancing time of a work measurement system. That is why the MOST Work Measurement Technique was designed to have a consistent balancing time of approximately 2 minutes. Also, within the MOST Application Systems, the analyst has a choice of time standards balancing times which range from ±5% at 8 hours to ±10% for 4 weeks.

MTM-1, MTM-2, or MOST?

Now that we know the theory behind the construction of MOST, which of these three systems, MTM-1, MTM-2, or MOST, should be used in a work measurement situation? On the one hand, we have perhaps the most precise predetermined motion time system available in MTM-1, and on the other, the most economical in MOST. Lying somewhere in the middle there is MTM-2. Which of these three techniques is the most appropriate for any given situation?

With respect to accuracy, the generally accepted standard for industry is ±5% at a 95% confidence level. Selection of the appropriate technique to be used is, therefore, based on the most economical system meeting this criterion. As we have seen earlier in this chapter, the point at which this accuracy criterion is achieved is determined by the system's balancing time. Balancing time, then, has a basic influence on the selection of the most appropriate work measurement technique to be used.

MOST was constructed to have a consistent balancing time of around 2 minutes. With MTM-1 and MTM-2, balancing time was not a factor in the system design. Nonetheless, each of the MTM systems possesses a balancing time of its own, which can be determined by statistical analysis. The previously mentioned study by Dr. Brinckloe concluded that the relative balancing times (based on ± 5% accuracy at a 95% confidence level) for the three systems are as follows:

MTM-1, 600 TMU (average)
MTM-2, 1600 TMU (average)

Table A.4 Example of Balancing Effect

	TRUE TIME IN HOURS	ALLOWED TIME IN HOURS	ACTUAL DEVIATION	
			± HOURS	± %
BALANCING TIME ↓	5.3	4.9	− .4	− 7.5
	3.9	3.6	− .3	− 7.7
	8.0	8.1	+ .1	+ 1.3
	2.0	1.6	− .4	− 20.0
	7.8	8.1	+ .3	+ 3.8
	1.4	1.6	+ .2	+ 14.3
	5.5	4.9	− .6	− 10.9
	3.3	3.6	+ .3	+ 9.1
	4.8	4.9	+ .1	+ 2.1
	42.0	41.3	− .7	− 1.7 %

MOST, 3235 TMU
MTM-3, 16,000 TMU*

So, it would seem that the analyst need only select the appropriate system based on the cycle time of the activity to be measured. That is, activities with cycle times of 600 TMU or more may be measured with MTM-1, those of 1600 TMU or more measured with MTM-2, and those 3200 TMU or more with MOST. Notice that for cycle times less than 600 TMU, the accuracy of ±5% cannot be assured, even when using MTM-1.

Theoretically correct, the activity's cycle time can be used as a general guideline toward proper system selection. Practically, however, other factors must be considered in order to fully realize the capabilities of each system. Does each activity cycle repeat identically from one occurrence to the next? If each cycle is not identical, variations are present which will tend to be leveled out due to the balancing effect.

Also worth noting is the fact that we must deal with averages in most work measurement situations. That is, average distances, average weight, average types of motion, and so on. Consider for example, this situation. Seated at a punch press, an operator gets a part that has just been formed and moves it to one of 12 spaces in a parts tray within reach to the right. The distance from the press to the spaces on the tray varies from 8 to 32 in. (20–81 cm). If that were the entire operaton to be measured, which work measurement technique *would* you

*Not included in study. Figure is given for comparison only.

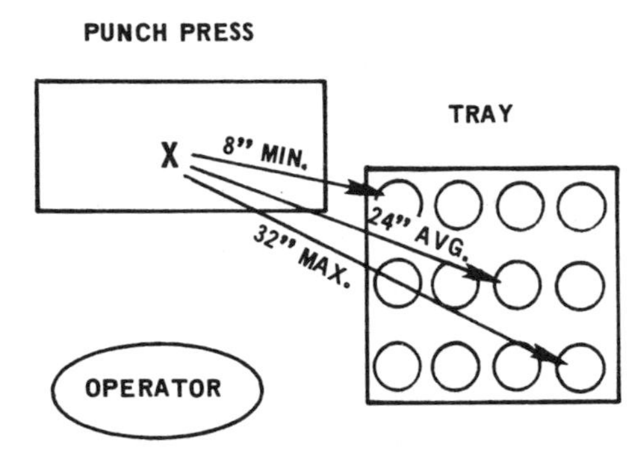

Figure A.4 Punch press operation.

choose—MTM-1, MTM-2, or MOST? Which technique *could* you choose? The cycle time is obviously very short. To achieve an accurate analysis, your choice should be a very detailed system. Or should it?

An MTM-1 analysis of the punch press operation (Figure A.4) might be:

MTM-1		TMU
R24A		14.9
G1A		2.0
M24B		20.6
RL1		2.0
	Total	39.5

Notice that in this case the MTM-1 analyst assumed an average distance of 24 in. (61 cm) for the operation (an actual case). In actuality, however, the distance varied from a minimum of 8 in. (20 cm) to a maximum of 32 in. (81 cm).

MTM-1 Minimum (8 in., 20 cm)		TMU	MTM-1 Maximum (32 in., 81 cm)		TMU
R8A		7.9	R32A		18.3
G1A		2.0	G1A		2.0
M8B		10.6	M32B		25.5
RL1		2.0	RL1		2.0
	Total	22.5		Total	47.8

Deviations

Total deviation (8-32 in., 20-81 cm), 64%
Minimum to average (8-24 in., 20-61 cm), 43%
Average to Maximum (24-32 in., 61-81 cm), 21%

Since variation did exist, the MTM-1 analyst was quite correct in using an average distance in the analysis.

An MTM-2 analysis of the same operation will give:

MTM-2		TMU
GB32		23
PA32		20
	Total	43

The corresponding MOST analysis using one General Move Sequence Model is:

MOST	TMU
$A_1\ B_0\ G_1\ A_1\ B_0\ P_1\ A_0$	40

All three time values for the operation (39.5, 43, and 40 TMU) are based on averages.

With MTM-1, the analyst selected an average distance for the reach and move motions based on a subjective judgement. MTM-2 time values are determined from the weighted average of different MTM-1 motion patterns. Index values in MOST are based on statistically calculated averages. The question no longer is "which is the correct analysis?" but "which is the most acceptable average?" No one can say with certainty which average is better. Therefore, when dealing with situations where variations in the operation occur from cycle to cycle, MOST gives results that are as good as the more detailed systems. In the analysis of an operation of any cycle length that contains variations, which measurement technique *could* be chosen? Any of them. MTM-1, MTM-2, and MOST will all produce an acceptable time value from an accuracy standpoint.*

Since almost every operation or manual activity performed in practical life contains method variations, the main purpose is not to try to analyze and measure one or all of the methods. Instead, it is a matter of controlling the variations out from a description of a representative or "average" method. In designing MOST, this fact was a major consideration. It became an important and fundamental principle in the buildup of the system. The index scale applied for MOST is no more than a series of statistically calculated averages, designed with the objective of absorbing method variations. Therefore, if the variations become larger than what is "allowed" for the actual interval, two or more time values must be applied for the operation. For example, apply a G_3 and a frequency to account for those times when one part is "interlocked" with another.

After having accepted the fact that methods variations do always occur (however, to a lesser extent for short-cycle jobs), the prime goal must be to establish a "good average" standard time value and to deal with inevitable variations in a controlled manner. Such a procedure results in a high degree of consistency.

*The practical accuracy of MTM-based standards is discussed in *The Impact of Variation in Method or Workplace on the System Precision of MTM Based Standards,* University Research Institute, March 1979.

In order to use a detailed work measurement system, like MTM-1 or MTM-2 made up of a large number of more or less independent elements, a great deal of subjectivity will be required in making decisions for an analysis. "Subjective averaging" can be good or bad. One thing is certain: It is not a consistent method, and the results will be greatly influenced by the individual's experience and performance. MOST in this respect is definitely more objective and consistent because the averages have been statistically established and can be consistently applied.

The situation above points out several key factors that must be considered in choosing a work measurement technique:

Variation

When analyzing an operation that varies from cycle to cycle, even the most detailed systems concede accuracy to the analytical technique of averaging. The question is then a subjective one of choosing the average that appears to best fit the situation.

System Accuracy

When will the technique being considered reach the accuracy level of ±5% with a 95% confidence level? The "balancing time" of the system is an indication of the accuracy of that system; i.e., the shorter the balancing time the more accurate (detailed) the technique. An operation such as the punch press example is shorter than the balancing time for MTM-1. Therefore, the accuracy of the analysis will be somewhat less than ±5% at 95% confidence.

Theoretically, there is a difference in the balancing time between these systems mentioned above. Practically, this difference is almost negligible, mainly due to the inability to fully and properly utilize that level of detail on which the highly detailed systems' balancing times are calculated.

Application Error

More detailed systems, with their inherent greater numbers of variables and finer discrimination of distances, and so on, afford the analyst greater opportunity to assign incorrect values to, or omit basic motions from the calculation. MOST forces the operator, through preprinted motion sequence models, to consider all the variables. Also, the analyst is taxed with fewer decisions to make, and the elements available for selection are fewer in MOST. As a result, the analyst has the opportunity to make fewer errors. Therefore, we can say that the accuracy MOST concedes through its larger balancing time could be compensated for by reduced applicator error.

Consistency

Through the use of the sequence model approach, analysts are aided in making the correct decisions. The result is smaller deviations among analysts compared to other predetermined motion time systems. As stated above, the statistically derived index scale will further emphasize the consistency of MOST.

Conclusion

The factors mentioned above play an important role in the choice of a technique to use in analyzing a short-cycle operation. MOST is a predetermined motion time system that can be employed with confidence for an operation of any length where variation exists from cycle to cycle. MOST, however, should not be used for highly repetitive operations where little or no variation exists between cycles or for identical operations.

In industry, variations are prevalent in a majority of operations. Therefore, *MOST can be used* for the vast majority of activities and operations *independent of cycle length or degree of repetitiveness.* From a practical standpoint and properly applied, MOST will produce the required accuracy and so will MTM-1 and MTM-2. There is no reason to believe any one system will lead to more accurate results than any other.

The only exception is when identical motion patterns occur. These are normally of very short duration: a few seconds, 10 to 15 at the most. Longer cycles can hardly be performed identically. Therefore, MTM-1 should be used for *identical* motion patterns independent of cycle length.

Even though the system precision is somewhat higher for MTM-1 and MTM-2, MOST can comfortably make up the difference through reduced applicator deviations and consistency. Consequently, the three systems should be equal regarding total overall accuracy.

Adequate accuracy, combined with high application speed and appealing simplicity, should make MOST the logical and natural choice for any work measurement situation, with the exception of completely identical motion patterns.

appendix B
MOST Calculation Examples

MOST-calculation

Code	01 05 0201
Date	3/18/75
Sign.	K.Z.
Page	1/1

Area: MACHINE SHOP TURRET LATHE

Activity: ADVANCE AND RETRACT DRILL IN TURRET TO WORKPIECE

Conditions:

No.	Method
1	INDEX TURRET
2	MOVE DRILL (TURRET) TO WORKPIECE
3	ADJUST COOLANT JET TO WORKPIECE
4	START AND STOP FEED
5	CHANGE RPM
6	RETRACT DRILL (TURRET) FROM WORKPIECE

No.	Sequence Model	Fr	TMU
	A B G A B P A		
	A B G A B P A		
	A B G A B P A		
	A B G A B P A		
	A B G A B P A		
	A B G A B P A		
	A B G A B P A		
	A B G A B P A		
	A B G A B P A		
	A B G A B P A		
	A B G A B P A		
	A B G A B P A		
	A B G A B P A		
	A B G A B P A		
	A B G A B P A		
	A B G A B P A		
	A B G A B P A		
	A B G A B P A		
1	$A_1 B_0 G_1 M_3 X_0 I_0 A_0$		50
2	$A_1 B_0 G_1 M_6 X_0 I_3 A_0$		110
3	$A_1 B_0 G_1 M_1 X_0 I_1 A_0$		40
4	$A_1 B_0 G_1 M_1 X_0 I_0 A_0$	2	60
5	$A_1 B_0 G_1 M_3 X_0 I_0 A_0$	2	100
6	$A_1 B_0 G_1 M_6 X_0 I_0 A_0$	2	80
	A B G M X I A		
	A B G M X I A		
	A B G A B P A B P A		
	A B G A B P A B P A		
	A B G A B P A B P A		
	A B G A B P A B P A		
	A B G A B P A B P A		
	A B G A B P A B P A		
	A B G A B P A B P A		
	A B G A B P A B P A		
	A B G A B P A B P A		
	A B G A B P A B P A		
	A B G A B P A B P A		
	A B G A B P A B P A		
	A B G A B P A B P A		
	A B G A B P A B P A		

TIME = .26 ~~millihours (mh.)~~/minutes (min.) 440

MOST-calculation	
Code	1 09 03 05 3 02
Date	11/13/72
Sign.	A.A.
Page	1/1

Area: MACHINE SHOP. VERTICAL LATHE

Activity: CHANGE WORKPIECE ON FACEPLATE WITH JIB CRANE

Conditions:

No.	Method
1	GET LIFT STRAPS
2	GET CRANE HOOK
3	USING JIB CRANE. REMOVE WORKPIECE FROM LATHE AND PLACE ON WORKBENCH
4	USING JIB CRANE. MOVE NEW WORKPIECE TO LATHE
5	ASIDE LIFT STRAPS
6	ASIDE CRANE HOOK
7	CLEAN WORKPIECE WITH AIR HOSE

No.	Sequence Model	Fr	TMU
1	$A_{10}\ B_{6}\ G_{3}\ A_{10}\ B_{0}\ P_{0}\ A_{0}$	1/2	145
2	$A_{10}\ B_{6}\ G_{3}\ A_{10}\ B_{0}\ P_{0}\ A_{0}$	1/2	145
5	$A_{1}\ B_{6}\ G_{3}\ A_{10}\ B_{0}\ P_{3}\ A_{10}$	1/2	165
6	$A_{1}\ B_{0}\ G_{3}\ A_{10}\ B_{0}\ P_{3}\ A_{10}$	1/2	135
	A B G A B P A		
	A B G A B P A		
	A B G A B P A		
	A B G A B P A		
	A B G A B P A		
	A B G A B P A		
	A B G A B P A		
	A B G A B P A		
	A B G A B P A		
	A B G A B P A		
	A B G A B P A		
	A B G A B P A		
	A B G A B P A		
	A B G A B P A		
	A B G M X I A		
	A B G M X I A		
	A B G M X I A		
	A B G M X I A		
	A B G M X I A		
	A B G M X I A		
	A B G M X I A		
	A B G M X I A		
7	$A_{3}\ B_{0}\ G_{1}\ A_{3}\ B_{0}\ P_{1}\ S_{32}\ A_{3}\ B_{0}\ P_{3}\ A_{0}$	2	920
	A B G A B P A B P A		
	A B G A B P A B P A		
	A B G A B P A B P A		
	A B G A B P A B P A		
	A B G A B P A B P A		
	A B G A B P A B P A		
	A B G A B P A B P A		
	A B G A B P A B P A		
	A B G A B P A B P A		
	A B G A B P A B P A		
	A B G A B P A B P A		
	A B G A B P A B P A		
	A B G A B P A B P A		
3	$A_{16}\ T_{54}\ K_{220}\ F_{24}\ V_{32}\ L_{81}\ V_{10}\ P_{24}\ T_{42}\ A_{0}$		5030
4	$A_{16}\ T_{54}\ K_{220}\ F_{24}\ V_{10}\ L_{81}\ V_{32}\ P_{16}\ T_{42}\ A_{10}$		5050

TIME = 6.9 ~~millihours (mh.)~~/minutes (min.) | 11590

MOST-calculation

Code	1 1 1 0 4 1 0 2 0 6
Date	11/28/77
Sign.	WMY
Page	1/1

Area MACHINING - ENGINE LATHE

Activity CHANGE TOOL IN DRILL CHUCK

Conditions JACOBS CHUCK, TOOL ON SHELF OR RACK

No.	Method
1	LOOSEN CHUCK WITH CHUCK WRENCH
2	OPEN CHUCK JAWS WITH FINGERS
3	REMOVE TOOL AND LAY ASIDE
4	OBTAIN TOOL AND POSITION IN CHUCK
5	HAND TIGHTEN CHUCK
6	TIGHTEN CHUCK WITH CHUCK WRENCH

No.	Sequence Model	Fr	TMU
3	$A_1 B_0 G_1 A_6 B_0 P_1 A_0$		90
4	$A_3 B_0 G_1 A_6 B_0 P_3 A_1$		140
1	$A_1 B_0 G_1 A_1 B_0 P_3 L_{16} A_1 B_0 P_1 A_0$		240
2	$A_0 B_0 G_0 A_1 B_0 P_1 L_3 A_0 B_0 P_0 A_0$		50
5	$A_0 B_0 G_0 A_1 B_0 P_1 F_3 A_0 B_0 P_0 A_0$		50
6	$A_1 B_0 G_1 A_1 B_0 P_3 F_{16} A_1 B_0 P_1 A_0$		240

TIME = 8.1 millihours (mh.) 810

MOST-calculation

Code	1030310201
Date	2/10/75
Sign.	KZ
Page	1/1

Area: MACHINING

Activity: CHANGE WORKPIECE IN 3-JAW CHUCK WITH T-WRENCH AT ENGINE LATHE

Conditions:

No.	Method
1	LOOSEN WORKPIECE WITH T-WRENCH
2	MOVE WORKPIECE FROM 3-JAW CHUCK TO PALLET 1
3	PLACE WORKPIECE FROM PALLET 2 TO 3-JAW CHUCK
4	FASTEN WORKPIECE IN 3-JAW CHUCK WITH T-WRENCH
5	PULL LEVER ON LATHE TO START SPINDLE FOR CENTRICITY CHECK, PT. 1 SEC

No.	Sequence Model	Fr	TMU
2	$A_1\ B_0\ G_1\ A_6\ B_6\ P_1\ A_0$		150
3	$A_3\ B_6\ G_1\ A_{10}\ B_0\ P_3\ A_0$		230
	A B G A B P A		
	A B G A B P A		
	A B G A B P A		
	A B G A B P A		
	A B G A B P A		
	A B G A B P A		
	A B G A B P A		
	A B G A B P A		
	A B G A B P A		
	A B G A B P A		
	A B G A B P A		
	A B G A B P A		
	A B G A B P A		
	A B G A B P A		
5	$A_1\ B_0\ G_1\ M_3\ X_3\ I_0\ A_0$		80
	A B G M X I A		
	A B G M X I A		
	A B G M X I A		
	A B G M X I A		
	A B G M X I A		
	A B G M X I A		
1	$A_1\ B_0\ G_1\ A_1\ B_0\ P_3\ L_{16}\ A_1\ B_0\ P_1\ A_0$		240
4	$A_1\ B_0\ G_1\ A_1\ B_0\ P_3\ F_{16}\ A_1\ B_0\ P_1\ A_0$		240
	A B G A B P A B P A		
	A B G A B P A B P A		
	A B G A B P A B P A		
	A B G A B P A B P A		
	A B G A B P A B P A		
	A B G A B P A B P A		
	A B G A B P A B P A		
	A B G A B P A B P A		
	A B G A B P A B P A		
	A B G A B P A B P A		

TIME = .564 minutes (min.) **940**

MOST-calculation

Code	3000510301
Date	9/29/76
Sign.	KZ
Page	1/1

Area: 4-SPINDLE GANG DRILL

Activity: DRILL HOLE IN DRIVE GEAR

Conditions:

No.	Method
1	PLACE PART IN FIXTURE
2	PUSH LEVER TO HOLD PART
3	APPROACH SPINDLE TO PART
4	START DRILLING OPERATION
5	PLACE PARTS ON MACHINE TABLE, SIMO
6	PULL LEVER ON FIXTURE TO LOOSEN PART
7	REMOVE PART TO TOTE BOX
8	CLEAN FIXTURE W/AIR

No.	Sequence Model	Fr	TMU
1	$A_1 B_0 G_3 A_1 B_0 P_3 A_0$		80
5	$A_3 B_6 G_3 A_3 B_0 P_3 A_0$	1/4	(45)
7	$(A_1 B_0 G_1) A_1 B_0 P_1 A_0$		20
2	$A_1 B_0 G_1 M_1 X_0 I_0 A_0$		30
3	$A_1 B_0 G_1 M_{10} X_0 I_0 A_0$		120
4	$(A_1) B_0 G_1 M_1 X_{113} I_0 A_0$		1150
6	$A_1 B_0 G_1 M_1 X_0 I_0 A_0$		30
8	$A_1 B_0 G_1 A_1 B_0 P_1 S_6 A_1 B_0 P_1 A_0$		120

TIME = .93 minutes (min.) — 1,550

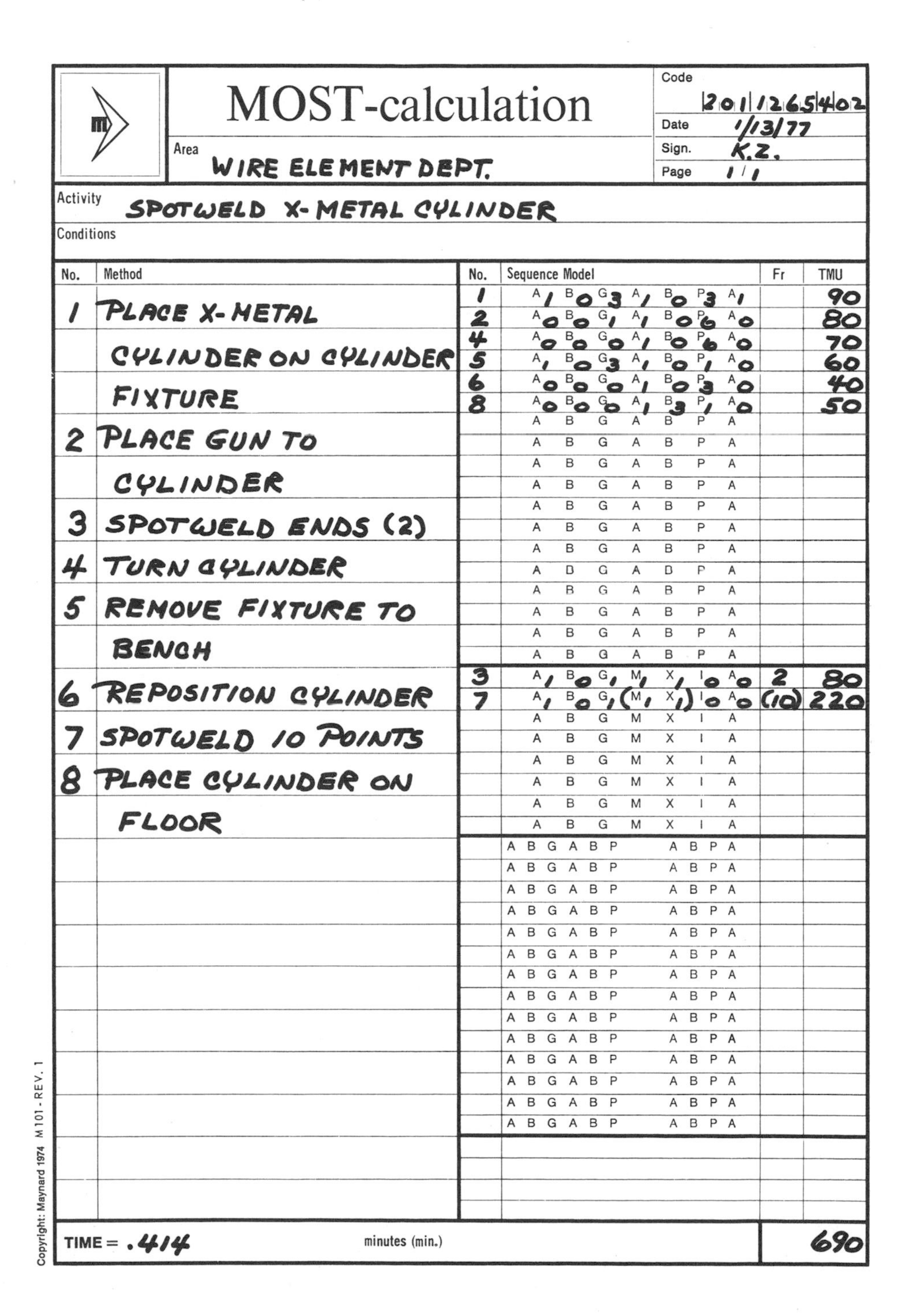

MOST-calculation

Code: 2011265402
Date: 1/13/77
Sign.: K.Z.
Page: 1/1

Area: WIRE ELEMENT DEPT.

Activity: SPOTWELD X-METAL CYLINDER

Conditions:

No.	Method
1	PLACE X-METAL CYLINDER ON CYLINDER FIXTURE
2	PLACE GUN TO CYLINDER
3	SPOTWELD ENDS (2)
4	TURN CYLINDER
5	REMOVE FIXTURE TO BENCH
6	REPOSITION CYLINDER
7	SPOTWELD 10 POINTS
8	PLACE CYLINDER ON FLOOR

No.	Sequence Model	Fr	TMU
1	$A_1\ B_0\ G_3\ A_1\ B_0\ P_3\ A_1$		90
2	$A_0\ B_0\ G_1\ A_1\ B_0\ P_6\ A_0$		80
4	$A_0\ B_0\ G_0\ A_1\ B_0\ P_6\ A_0$		70
5	$A_1\ B_0\ G_3\ A_1\ B_0\ P_1\ A_0$		60
6	$A_0\ B_0\ G_0\ A_1\ B_0\ P_3\ A_0$		40
8	$A_0\ B_0\ G_0\ A_1\ B_3\ P_1\ A_0$		50
3	$A_1\ B_0\ G_1\ M_1\ X_1\ I_0\ A_0$	2	80
7	$A_1\ B_0\ G_1\ (M_1\ X_1)\ I_0\ A_0$	(10)	220

TIME = .414 minutes (min.) | 690

MOST-calculation

Code: 402600 2201
Date: 3/19/75
Sign.: K.Z.
Page: 1/1

Area: FABRICATION

Activity: PUNCH HOLES IN PLATE AT STRIPPIT PUNCH PRESS, PER 2-4 HOLES, PLATE SIZE = 24 X 30"

Conditions:

No.	Method
1	POSITION PLATE FROM HAND-TRUCK TO PUNCH PRESS
2	PUSH BUTTONS FOR PUNCHING HOLES F_3
3	SLIDE PLATE AGAINST STOP AT PRESS F_2
4	MOVE PLATE FROM PRESS TO HANDTRUCK 2
5	CHECK HOLE LOCATIONS WITH CALIPER AND RETURN TO HANDTRUCK 2, F 1/4
6	SLIDE PLATE AT HANDTRUCK 2, F 1/4

No.	Sequence Model	Fr	TMU
1	$A_3\ B_0\ G_1\ A_3\ B_0\ P_6\ A_0$		130
4	$A_1\ B_0\ G_1\ A_6\ B_0\ P_1\ A_6$		150
	A B G A B P A		
	A B G A B P A		
	A B G A B P A		
	A B G A B P A		
	A B G A B P A		
	A B G A B P A		
	A B G A B P A		
	A B G A B P A		
	A B G A B P A		
	A B G A B P A		
	A B G A B P A		
	A B G A B P A		
	A B G A B P A		
	A B G A B P A		
	A B G A B P A		
2	$A_1\ B_0\ G_1\ M_1\ X_1\ I_0\ A_0$	3	120
3	$A_1\ B_0\ G_1\ M_3\ X_0\ I_0\ A_0$	2	100
6	$A_1\ B_0\ G_1\ M_3\ X_0\ I_0\ A_0$	1/4	13
	A B G M X I A		
	A B G M X I A		
	A B G M X I A		
	A B G M X I A		
	A B G M X I A		
5	$A_3\ B_0\ G_1\ A_3\ B_0\ P_1\ M_{24}\ A_3\ B_0\ P_1\ A_3$	1/4	98
	A B G A B P A B P A		
	A B G A B P A B P A		
	A B G A B P A B P A		
	A B G A B P A B P A		
	A B G A B P A B P A		
	A B G A B P A B P A		
	A B G A B P A B P A		
	A B G A B P A B P A		
	A B G A B P A B P A		
	A B G A B P A B P A		
	A B G A B P A B P A		
	A B G A B P A B P A		
	A B G A B P A B P A		

TIME = .367 minutes (min.) | 611

MOST-calculation

Code 007|8362|2|03
Date 3/19/75
Sign. K.Z.
Page 1/1

Area ASSEMBLY

Activity ASSEMBLE RESISTOR OR DIODE ON PC-BOARD

Conditions

No.	Method
1	TURN CAROUSEL TRAY WITH COMPONENTS F-1/8
2	GET AND MOVE 4 COMPONENTS FROM TRAY BIN TO BENCH, F-1/4
3	PICK UP COMPONENT
4	BEND LEGS ON COMPONENT WITH PLIERS
5	MOVE COMPONENT TO BENCH
6	READ DRAWING, 10 DIGITS, TO LOCATE COMPONENT TO PC-BOARD
7	POSITION COMPONENT TO PC-BOARD
8	FASTEN COMPONENT LEGS WITH PLIERS
9	CUTOFF EXCESS WIRE WITH DIKE

No.	Sequence Model	Fr	TMU
2	$A_1 B_0 G_3 A_1 B_0 P_1 A_0$	1/4	15
3	$A_1 B_0 G_1 A_1 B_0 P_0 A_0$		30
5	$A_0 B_0 G_0 A_1 B_0 P_1 A_0$		20
7	$A_1 B_0 G_1 A_1 B_0 P_6 A_0$		90
	A B G A B P A		
	A B G A B P A		
	A B G A B P A		
	A B G A B P A		
	A B G A B P A		
	A B G A B P A		
	A B G A B P A		
	A B G A B P A		
	A B G A B P A		
	A B G A B P A		
	A B G A B P A		
	A B G A B P A		
	A B G A B P A		
	A B G A B P A		
1	$A_1 B_0 G_1 M_3 X_0 I_3 A_0$	1/8	10
	A B G M X I A		
	A B G M X I A		
	A B G M X I A		
	A B G M X I A		
	A B G M X I A		
	A B G M X I A		
	A B G M X I A		
4	$A_1 B_0 G_1 A_1 B_0 (P_3 A_0 C_1) A_1 B_0 P_1 A_0$ (8)	1/4	92
6	$A_0 B_0 G_0 A_0 B_0 P_0 T_{10} A_0 B_0 P_0 A_0$		100
8	$A_1 B_0 G_1 A_1 B_0 (P_1 A_0 F_3) A_1 B_0 P_1 A_0$ (2)		130
9	$A_1 B_0 G_1 A_1 B_0 (P_3 A_0 C_1) A_1 B_0 P_1 A_0$ (8)	1/4	92
	A B G A B P A B P A		
	A B G A B P A B P A		
	A B G A B P A B P A		
	A B G A B P A B P A		
	A B G A B P A B P A		
	A B G A B P A B P A		
	A B G A B P A B P A		
	A B G A B P A B P A		
	A B G A B P A B P A		
	A B G A B P A B P A		

TIME = .347 minutes (min.) 579

MOST-calculation

Code	1 01 5373 302
Date	3/25/77
Sign.	S
Page	1/1

Area: AUTO ASSEMBLY

Activity: ASSEMBLE 4 SHOCKABSORBERS (FRONT) TO FRAME

Conditions:

No.	Method
1	GET AND PICK UP 4 SHOCK-ABSORBERS
2	HOLD AND PLACE 3 SHOCKABSORBERS TO FRAME
3	LOOSEN 4 NUTS 5 SPINS with FINGERS
4	PICK UP RETAINER AND BUSHING FROM SHOCK-ABSORBER
5	HOLD AND PLACE SHOCK-ABSORBER IN BRACKET
6	HOLD AND PLACE BUSHING AND RETAINER TO SHOCKABSORBERS
7	HOLD AND FASTEN 4 NUTS 5 SPINS WITH FINGERS TO SHOCKABSORBER
8	PICK UP SHOCKABSORBERS
9	GET AND PICK UP 2 SHOCKABS
10	HOLD AND MOVE SHOCK-ABSORBERS TO FRAME

No.	Sequence Model	Fr	TMU
1	$(A_1\ B_0\ G_3)\ A_6\ B_0\ P_0\ A_0$	(2)	140
2	$A_0\ B_0\ G_0\ A_1\ B_0\ P_3\ A_0$	3	120
4	$A_1\ B_0\ G_1\ A_1\ B_0\ P_0\ A_0$		30
5	$A_0\ B_0\ G_0\ A_1\ B_6\ P_3\ A_0$	4	400
6	$A_0\ B_0\ G_0\ A_1\ B_0\ P_3\ A_0$	4	160
8	$A_1\ B_0\ G_1\ A_1\ B_0\ P_0\ A_0$		30
9	$A_1\ B_0\ G_3\ A_{16}\ B_0\ P_0\ A_0$		200
10	$A_0\ B_0\ G_0\ A_1\ B_0\ P_1\ A_0$		20
	A B G A B P A		
	A B G A B P A		
	A B G A B P A		
	A B G A B P A		
	A B G A B P A		
	A B G A B P A		
	A B G A B P A		
	A B G A B P A		
	A B G A B P A		
	A B G A B P A		
	A B G M X I A		
	A B G M X I A		
	A B G M X I A		
	A B G M X I A		
	A B G M X I A		
	A B G M X I A		
	A B G M X I A		
	A B G M X I A		
3	$A_0\ B_0\ G_0\ A_1\ B_0\ P_1\ L_{10}\ A_0\ B_0\ P_0\ A_0$	4	480
7	$A_0\ B_0\ G_0\ A_1\ B_0\ P_3\ F_{10}\ A_0\ B_0\ P_0\ A_0$	4	560
	A B G A B P A B P A		
	A B G A B P A B P A		
	A B G A B P A B P A		
	A B G A B P A B P A		
	A B G A B P A B P A		
	A B G A B P A B P A		
	A B G A B P A B P A		
	A B G A B P A B P A		
	A B G A B P A B P A		
	A B G A B P A B P A		
	A B G A B P A B P A		
	A B G A B P A B P A		

TIME = 1.28 minutes (min.) — 2140

MOST-calculation

Code	0050605201
Date	9/13/74
Sign.	K.Z.
Page	1/1

Area: ELECTRICAL MAINTENANCE

Activity: REPLACE LIGHT SWITCH

Conditions:

No.	Method
1	REMOVE COVER PLATE WITH SCREWDRIVER (2 SCREWS)
2	REMOVE SWITCH FROM BOX
3	REMOVE 2 SCREWS WITH SCREWDRIVER
4	DISCONNECT 2 WIRES
5	ASIDE SWITCH TO POCKET
6	STRAIGHTEN 2 WIRES
7	GET NEW SWITCH FROM POCKET TO WIRES
8	CONNECT 2 WIRES
9	POSITION SWITCH IN BOX
10	PLACE COVER PLATE
11	FASTEN COVER PLATE WITH SCREWDRIVER

No.	Sequence Model	Fr	TMU
2	A1 B0 G3 A1 B0 P1 A0		60
5	A1 B0 (G3) A1 B6 P1 A0	(2)	150
7	A1 B0 G3 A1 B0 P6 A0		110
9	A0 B0 G0 A1 B0 P6 A0		70
10	A1 B6 G3 A1 B0 P3 A0		140
	A B G A B P A		
	A B G A B P A		
	A B G A B P A		
	A B G A B P A		
	A B G A B P A		
	A B G A B P A		
	A B G A B P A		
	A B G A B P A		
	A B G A B P A		
	A B G A B P A		
	A B G A B P A		
	A B G A B P A		
	A B G A B P A		
	A B G M X I A		
	A B G M X I A		
	A B G M X I A		
	A B G M X I A		
	A B G M X I A		
	A B G M X I A		
	A B G M X I A		
	A B G M X I A		
1	A1 B0 G1 A1 B0 (P3 A0 L16) A1 B6 P1 A0	(2)	490
3	A0 B0 G0 A0 B0 (P3 A1 L54) A0 B0 P0 A0	(2)	1160
4	A0 B0 G0 A0 B0 (P3 A1 L10) A1 B0 P1 A0	(2)	300
6	A1 B0 G1 A0 B0 (P1 A1 F16) A1 B0 P1 A0	(2)	400
8	A1 B0 G1 A0 B0 (P3 A1 F16) A0 B0 P0 A0	(2)	420
11	A1 B0 G1 A0 B0 (P3 A1 F10) A1 B0 P1 A0	(2)	320
	A B G A B P A B P A		
	A B G A B P A B P A		
	A B G A B P A B P A		
	A B G A B P A B P A		
	A B G A B P A B P A		
	A B G A B P A B P A		
	A B G A B P A B P A		
	A B G A B P A B P A		

TIME = 2.2 ~~millihours (mhr)~~ / minutes (min.) — 3620

MOST-calculation

Code	5000001101
Date	6/20/77
Sign.	W.N.Y.
Page	1/1

Area: PIPE SHOP

Activity: BEND TUBE FOR CHAIR FRAME AT ROTARY COMPRESSION BENDING MACHINE (PER TUBE)

Conditions:

No.	Method
1	REMOVE TAPE FROM TUBES
2	ASIDE TAPE TO TRASH BIN AND RETURN TO WORKBENCH
3	GET TUBE FROM WORKBENCH AND PLACE IN BENDING MACHINE
4	PULL LEVER 1 WITH LEFT HAND (L.H.)
5	PULL SUPPORT WITH RIGHT HAND (R.H.)
6	PULL LEVER 2 WITH L.H.
7	PUSH ACTIVATING LEVER WITH R.H. TO BEND TUBE
8	PUSH LEVER 2 WITH L.H.
9	PUSH LEVER 1 WITH L.H.
10	REMOVE TUBE AND ASIDE TO LOWER SHELF
11	PULL ACTIVATING LEVER WITH R.H. TO RESET MACHINE (SIMO TO REMOVE TUBE)

No.	Sequence Model	Fr	TMU
1	(A_1 B_0 G_3 A_1 B_0 P_1) A_0 (5)	1/25	12
2	A_0 B_0 G_0 A_3 B_0 P_0 A_3	1/25	2
3	A_1 B_0 G_1 A_1 B_0 P_1 A_0		40
10	A_1 B_0 G_1 A_3 B_6 P_1 A_3		150
	A B G A B P A		
	A B G A B P A		
	A B G A B P A		
	A B G A B P A		
	A B G A B P A		
	A B G A B P A		
	A B G A B P A		
	A B G A B P A		
	A B G A B P A		
	A B G A B P A		
	A B G A B P A		
	A B G A B P A		
	A B G A B P A		
	A B G A B P A		
4	A_1 B_0 G_1 M_1 X_0 I_0 A_0		30
5	(A)$_1$ B_0 G_1 M_1 X_0 I_0 A_0		20
6	(A)$_1$ B_0 G_1 M_1 X_0 I_0 A_0		20
7	(A)$_1$ B_0 G_1 M_1 X_3 I_0 A_0		50
8	A_0 B_0 G_0 M_1 X_0 I_0 A_0		10
9	A_1 B_0 G_1 M_1 X_0 I_0 A_0		30
11	A_1 B_0 G_1 M_3 X_3 I_0 A_0		(80)
	A B G M X I A		
	A B G A B P A B P A		
	A B G A B P A B P A		
	A B G A B P A B P A		
	A B G A B P A B P A		
	A B G A B P A B P A		
	A B G A B P A B P A		
	A B G A B P A B P A		
	A B G A B P A B P A		
	A B G A B P A B P A		
	A B G A B P A B P A		
	A B G A B P A B P A		
	A B G A B P A B P A		
	A B G A B P A B P A		
	A B G A B P A B P A		

TIME = .22 minutes (min.) — 364

MOST-calculation

Code	0062070304
Date	11/17/76
Sign.	D. D.
Page	1/1

Area: YARN HAULER / CHECKER

Activity: UNLOAD EMPTY BOBBINS

Conditions: FROM BINS ON AUTOCONER

No.	Method
1	OBTAIN EMPTY TRUCK FROM STORAGE AREA, PUSH TO FRONT ALLEY OF AUTO-CONERS; INCLUDES RETURN AT END OF ACTIVITY.
2	PUSH TRUCK FROM BIN TO BIN IN ALLEY.
3	REMOVE BOBBIN BIN ON AUTOCONER.
4	LIFT BIN TO TRUCK.
5	DUMP BOBBINS INTO TRUCK.
6	REPLACE BIN ON AUTO-CONER.

No.	Sequence Model	Fr	TMU
4	$A_0 B_0 G_0 A_1 B_0 P_1 A_0$	10	200
6	$A_0 B_0 G_0 A_1 B_6 P_3 A_0$	10	1000
2	$A_1 B_0 G_3 M_3 X_0 I_0 A_{10}$	4	680
3	$A_1 B_6 G_1 M_3 X_0 I_0 A_0$	10	1100
5	$A_0 B_0 G_0 M_1 X_{10} I_0 A_0$	10	1100
1	$A_3 S_3 T_6 L_0 T_0 L_0 T_6 A_3$		2100

TIME = 3.71 minutes (min.) 6180

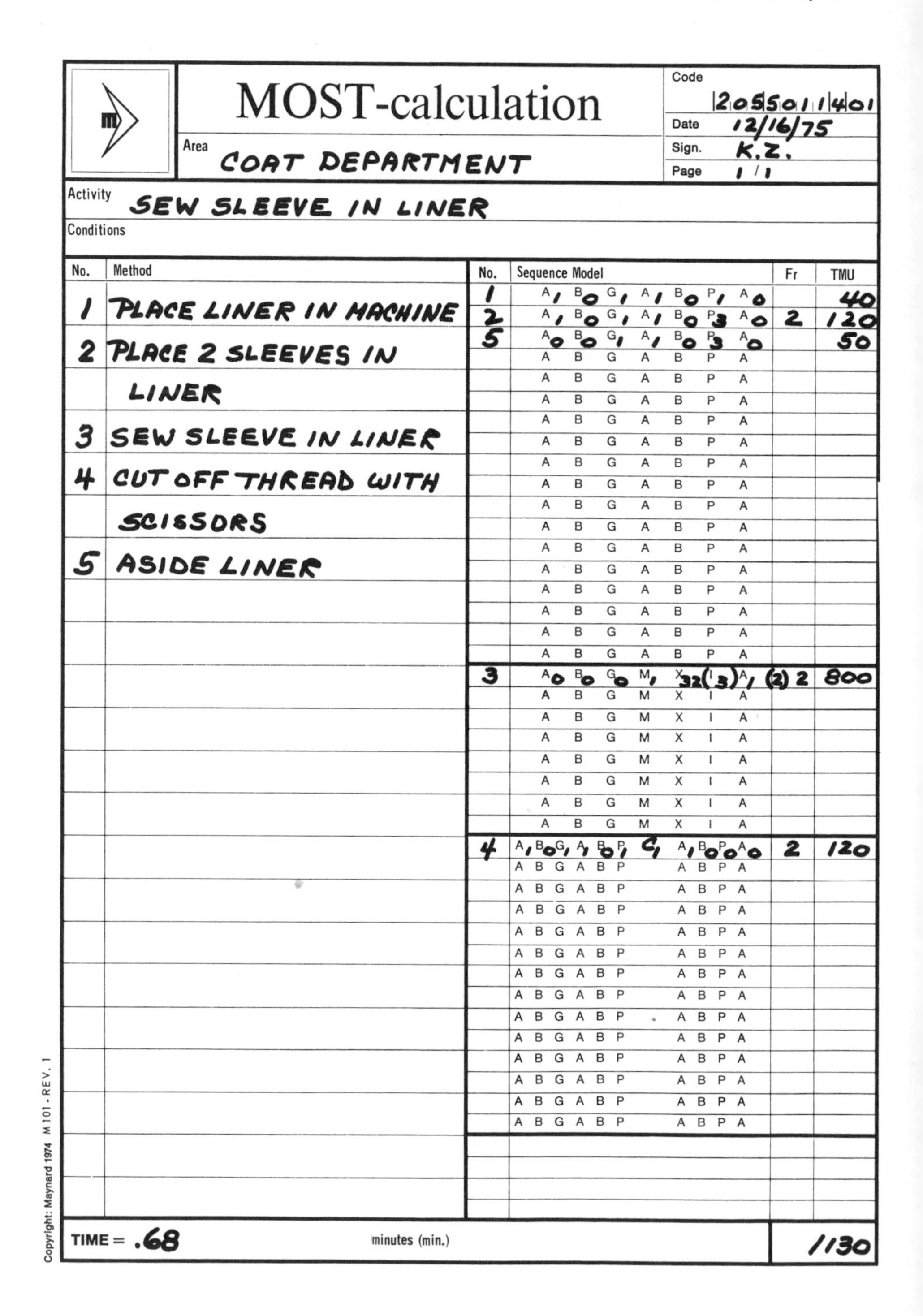

MOST-calculation

Code	2055011401
Date	12/16/75
Sign.	K.Z.
Page	1/1

Area: COAT DEPARTMENT

Activity: SEW SLEEVE IN LINER

Conditions

No.	Method
1	PLACE LINER IN MACHINE
2	PLACE 2 SLEEVES IN LINER
3	SEW SLEEVE IN LINER
4	CUT OFF THREAD WITH SCISSORS
5	ASIDE LINER

No.	Sequence Model	Fr	TMU
1	A_1 B_0 G_1 A_1 B_0 P_1 A_0		40
2	A_1 B_0 G_1 A_1 B_0 P_3 A_0	2	120
5	A_0 B_0 G_1 A_1 B_0 P_3 A_0		50
	A B G A B P A		
	A B G A B P A		
	A B G A B P A		
	A B G A B P A		
	A B G A B P A		
	A B G A B P A		
	A B G A B P A		
	A B G A B P A		
	A B G A B P A		
	A B G A B P A		
	A B G A B P A		
	A B G A B P A		
	A B G A B P A		
	A B G A B P A		
	A B G A B P A		
3	A_0 B_0 G_0 M_1 X_{32} (I_3) A_1 (2)	2	800
	A B G M X I A		
	A B G M X I A		
	A B G M X I A		
	A B G M X I A		
	A B G M X I A		
	A B G M X I A		
	A B G M X I A		
4	A_1 B_0 G_1 A_1 B_0 P_1 C_1 A_1 B_0 P_0 A_0	2	120
	A B G A B P A B P A		
	A B G A B P A B P A		
	A B G A B P A B P A		
	A B G A B P A B P A		
	A B G A B P A B P A		
	A B G A B P A B P A		
	A B G A B P A B P A		
	A B G A B P A B P A		
	A B G A B P A B P A		
	A B G A B P A B P A		
	A B G A B P A B P A		
	A B G A B P A B P A		

TIME = .68 minutes (min.) 1130

MOST-calculation

Area	APPAREL CUTTING
Code	1/185/09301
Date	9/23/76
Sign.	KZ
Page	1/1
Activity	HAND LAY UP 24 LAYERS HIGH (2 OPERATORS)
Conditions	

No.	Method
1	REMOVE CORE TUBE
2	PULL OUT TUBE
3	ASIDE TUBE
4	PLACE ROLL OF FABRIC ON MACHINE (LEFT END)
5	ADJUST CLAMPS (BOTH ENDS)
6	PLACE ROLL OF FABRIC ON MACHINE (RIGHT END)
7	PLACE END OF FABRIC
8	PLACE WEIGHTS TO HOLD END OF FABRIC
9	LAY UP FABRIC ON TABLE
10	ALIGN FABRIC EDGES (SIMO TO 9)
11	CUT OFF FABRIC W/SCISSORS
12	PUT WEIGHTS ON MACHINE
13	PLACE PATTERN PAPER ON TOP OF BATCH

No.	Sequence Model	Fr	TMU
1	$A_1\ B_0\ G_3\ A_1\ B_0\ P_0\ A_0$	2	100
3	$A_0\ B_0\ G_0\ A_1\ B_0\ P_1\ A_0$	2	40
4	$A_6\ B_6\ G_3\ A_1\ B_0\ P_3\ A_0$	2	380
6	$A_1\ B_0\ G_3\ A_6\ B_0\ P_3\ A_0$	2	260
7	$A_1\ B_0\ G_1\ A_1\ B_0\ P_6\ A_0$	24	2,160
8	$A_1\ B_0\ G_1\ A_1\ B_0\ P_1\ A_0$	48	1,920
10	$A_1\ B_0\ G_1\ A_0\ B_0\ P_1\ A_0$	48x5	(7,200)
12	$A_1\ B_0\ G_1\ A_1\ B_0\ P_1\ A_0$	2	80
13	$A_3\ B_0\ G_1\ A_3\ B_0\ P_3\ A_0$		100
	A B G A B P A		
	A B G A B P A		
	A B G A B P A		
	A B G A B P A		
	A B G A B P A		
	A B G A B P A		
	A B G A B P A		
	A B G A B P A		
	A B G A B P A		

No.	Sequence Model	Fr	TMU
2	$A_1\ B_0\ G_1\ M_3\ X_0\ I_0\ A_0$	2x2	200
5	$A_1\ B_0\ G_1\ M_3\ X_0\ I_0\ A_0$	2x2	200
9	$A_1\ B_0\ G_1\ M_3\ X_0\ I_0\ A_{16}$	24x2	10,080
	A B G M X I A		
	A B G M X I A		
	A B G M X I A		
	A B G M X I A		
	A B G M X I A		

No.	Sequence Model	Fr	TMU
11	$A_1\ B_0\ G_3\ A_1\ B_0\ P_1\ C_{10}\ A_1\ B_0\ P_3\ A_0$	25	5,000
	A B G A B P A B P A		
	A B G A B P A B P A		
	A B G A B P A B P A		
	A B G A B P A B P A		
	A B G A B P A B P A		
	A B G A B P A B P A		
	A B G A B P A B P A		
	A B G A B P A B P A		
	A B G A B P A B P A		
	A B G A B P A B P A		
	A B G A B P A B P A		
	A B G A B P A B P A		
	A B G A B P A B P A		

TIME = 12.31 minutes (min.) 20520

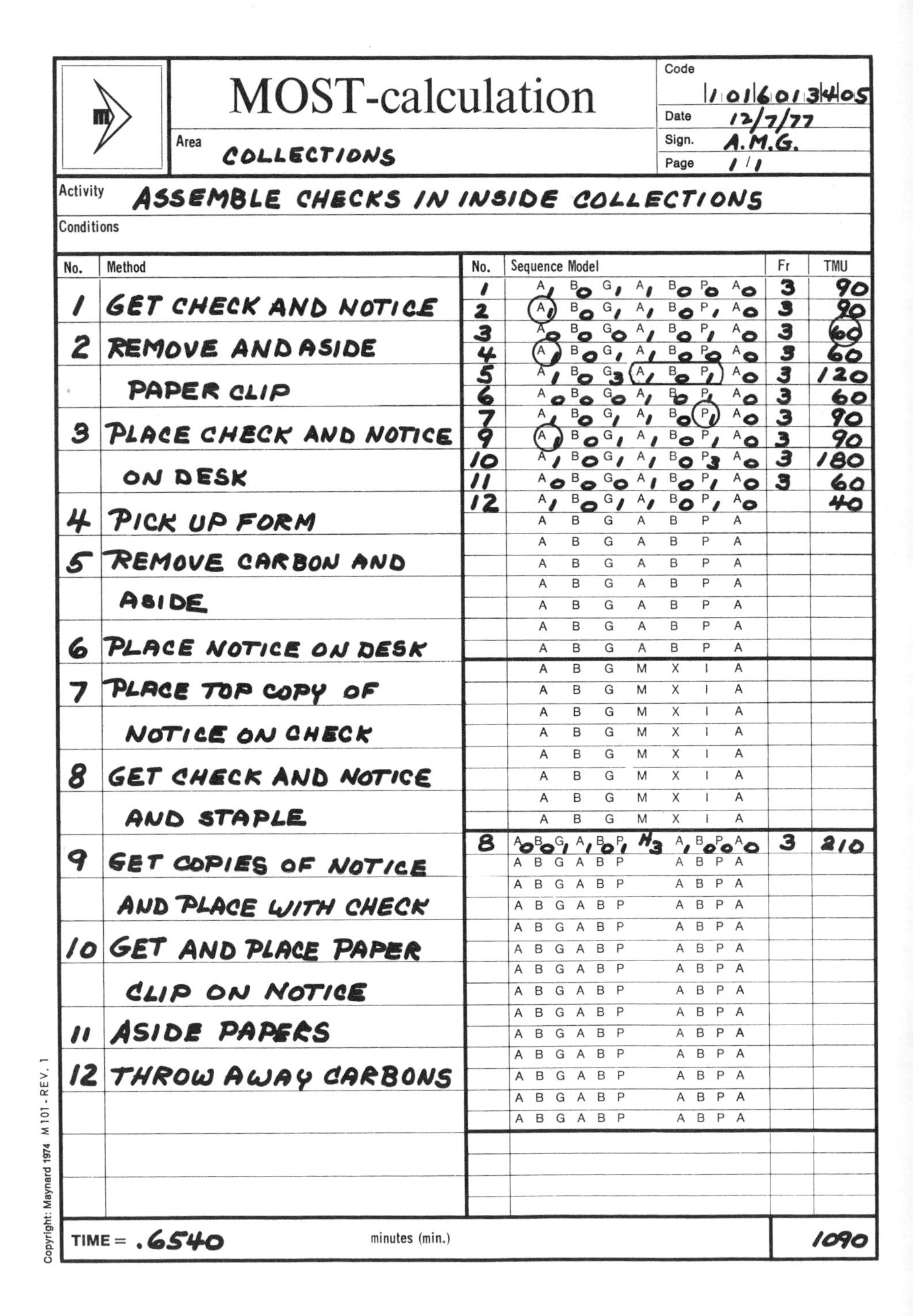

MOST-calculation

Code: 1 01 6 01 3 4 05
Date: 12/7/77
Sign.: A.M.G.
Page: 1/1

Area: COLLECTIONS

Activity: ASSEMBLE CHECKS IN INSIDE COLLECTIONS

Conditions:

No.	Method
1	GET CHECK AND NOTICE
2	REMOVE AND ASIDE PAPER CLIP
3	PLACE CHECK AND NOTICE ON DESK
4	PICK UP FORM
5	REMOVE CARBON AND ASIDE
6	PLACE NOTICE ON DESK
7	PLACE TOP COPY OF NOTICE ON CHECK
8	GET CHECK AND NOTICE AND STAPLE
9	GET COPIES OF NOTICE AND PLACE WITH CHECK
10	GET AND PLACE PAPER CLIP ON NOTICE
11	ASIDE PAPERS
12	THROW AWAY CARBONS

No.	Sequence Model	Fr	TMU
1	$A_1 B_0 G_1 A_1 B_0 P_0 A_0$	3	90
2	$A_1 B_0 G_1 A_1 B_0 P_1 A_0$	3	90
3	$A_0 B_0 G_0 A_1 B_0 P_1 A_0$	3	60
4	$A_1 B_0 G_1 A_1 B_0 P_0 A_0$	3	60
5	$A_1 B_0 G_3 (A_1 B_0 P_1) A_0$	3	120
6	$A_0 B_0 G_0 A_1 B_0 P_1 A_0$	3	60
7	$A_1 B_0 G_1 A_1 B_0 P_1 A_0$	3	90
9	$A_1 B_0 G_1 A_1 B_0 P_1 A_0$	3	90
10	$A_1 B_0 G_1 A_1 B_0 P_3 A_0$	3	180
11	$A_0 B_0 G_0 A_1 B_0 P_1 A_0$	3	60
12	$A_1 B_0 G_1 A_1 B_0 P_1 A_0$		40
	A B G A B P A		
	A B G A B P A		
	A B G A B P A		
	A B G A B P A		
	A B G A B P A		
	A B G A B P A		
	A B G A B P A		
	A B G M X I A		
	A B G M X I A		
	A B G M X I A		
	A B G M X I A		
	A B G M X I A		
	A B G M X I A		
	A B G M X I A		
	A B G M X I A		
8	$A_0 B_0 G_1 A_1 B_0 P_1$ H_3 $A_1 B_0 P_0 A_0$	3	210
	A B G A B P A B P A		
	A B G A B P A B P A		
	A B G A B P A B P A		
	A B G A B P A B P A		
	A B G A B P A B P A		
	A B G A B P A B P A		
	A B G A B P A B P A		
	A B G A B P A B P A		
	A B G A B P A B P A		
	A B G A B P A B P A		
	A B G A B P A B P A		
	A B G A B P A B P A		
	A B G A B P A B P A		

TIME = .6540 minutes (min.) 1090

MOST-calculation

Code	19046102301
Date	12/13/77
Sign.	B.B.
Page	1/1

Area: CASH AND MAIL

Activity: PREPARE DUPLICATE BILL FOR PAYMENT

Conditions:

No.	Method
1	OBTAIN PEN, WRITE DATE ON DUPLICATE BILL
2	COPY ACCOUNT NUMBER (11 DIGITS)
3	COPY SYMBOL FOR STATUS OF PAYMENT
4	COPY AMOUNT OF PAYMENT
5	COPY SYMBOL FOR TYPE OF PAYMENT
6	WRITE NAME OF CUSTOMER
7	WRITE SYMBOL ABOVE LAST NAME
8	WRITE ADDRESS OF CUSTOMER
9	WRITE CITY AND STATE OF CUSTOMER
10	WRITE INITIALS ON DUPLICATE BILL (2 LETTERS) AND ASIDE PEN

No.	Sequence Model	Fr	TMU
	A B G A B P A		
	A B G A B P A		
	A B G A B P A		
	A B G A B P A		
	A B G A B P A		
	A B G A B P A		
	A B G A B P A		
	A B G A B P A		
	A B G A B P A		
	A B G A B P A		
	A B G A B P A		
	A B G A B P A		
	A B G A B P A		
	A B G A B P A		
	A B G A B P A		
	A B G A B P A		
	A B G A B P A		
	A B G M X I A		
	A B G M X I A		
	A B G M X I A		
	A B G M X I A		
	A B G M X I A		
	A B G M X I A		
	A B G M X I A		
	A B G M X I A		
1	$A_1 B_0 G_1 A_1 B_0 P_1 R_{16} A_0 B_0 P_0 A_0$		200
2	$A_0 B_0 G_0 A_0 B_0 P_1 R_{24} A_0 B_0 P_0 A_0$		250
3	$A_0 B_0 G_0 A_0 B_0 P_1 R_3 A_0 B_0 P_0 A_0$		40
4	$A_0 B_0 G_0 A_0 B_0 P_1 R_6 A_0 B_0 P_0 A_0$		70
5	$A_0 B_0 G_0 A_0 B_0 P_1 R_3 A_0 B_0 P_0 A_0$		40
6	$A_0 B_0 G_0 A_1 B_0 P_1 R_{16} A_0 B_0 P_0 A_0$		180
7	$A_0 B_0 G_0 A_0 B_0 P_1 R_3 A_0 B_0 P_0 A_0$		40
8	$A_0 B_0 G_0 A_0 B_0 P_1 R_{32} A_0 B_0 P_0 A_0$		330
9	$A_0 B_0 G_0 A_0 B_0 P_1 R_{16} A_0 B_0 P_0 A_0$		170
10	$A_0 B_0 G_0 A_1 B_0 P_1 R_3 A_1 B_0 P_1 A_0$		70
	A B G A B P A B P A		
	A B G A B P A B P A		
	A B G A B P A B P A		
	A B G A B P A B P A		

TIME = .834 minutes (min.) 1390

Index